Food Engineering

Edited by Teodora Emilia Coldea

Published in London, United Kingdom

IntechOpen

Supporting open minds since 2005

Food Engineering
http://dx.doi.org/10.5772/intechopen.78442
Edited by Teodora Emilia Coldea

Contributors
Poliana Mendes De Souza, Miriam Aparecida De Aguilar Santos, Kássia Héllen Vieira, Regiane De Melo, Fabiana Regina Lima, Roberto Aguerre, Ethel Noemi Coscarello, Francisco Pantuso, María Laura Gómez Castro, Claudia Cecilia Larregain, Marko Petkovic, Roberto Jorge Aguerre, Cristiana Radulescu, Lavinia Claudia Buruleanu, Andreea Antonia Georgescu, Ioana Daniela Dulama, Romina Alina Vlaic, Crina Carmen Muresan, Sevastita Muste, Andruta Elena Muresan, Vlad Muresan, Georgiana Petrut, Mihaela Mihai, Ramona Suharoschi, Teodora Emilia Coldea

First published in London, United Kingdom, 2019 by IntechOpen
IntechOpen is the global imprint of INTECHOPEN LIMITED, registered in England and Wales, registration number: 11086078, The Shard, 25th floor, 32 London Bridge Street
London, SE19SG – United Kingdom
Printed in Croatia

British Library Cataloguing-in-Publication Data
A catalogue record for this book is available from the British Library

Additional hard and PDF copies can be obtained from orders@intechopen.com

Food Engineering
Edited by Teodora Emilia Coldea
p. cm.
Print ISBN 978-1-83881-975-0
Online ISBN 978-1-83881-976-7
eBook (PDF) ISBN 978-1-83881-977-4

For EU product safety concerns:
IN TECH d.o.o., Prolaz Marije Krucifikse Kozulić 3, 51000 Rijeka, Croatia, info@intechopen.com or visit our website at intechopen.com.

Meet the editor

Teodora Emilia Coldea was born in 1984. She graduated in Food Engineering from the University of Agricultural Sciences and Veterinary Medicine of Cluj-Napoca Romania in 2008, worked in industry for four years, and then returned to university. She completed her PhD in Biotechnology in 2011 and her postdoctoral studies in 2012. She joined the Faculty of Food Science and Technology in 2013 and is presently a lecturer teaching fermentation technology at the university. Since 2011, she has collaborated on projects in biotechnology, food engineering, product development, consumer preferences, and data analysis. She and her coauthors have presented more than 20 papers and posters at international conferences, and published over 30 papers in peer-reviewed journals and three international book chapters. Her past five years' research have focused the quality assessment (spectrometry, gas chromatography mass spectrometry, gas chromatography coupled with flame ionization detector, Fourier transform infrared spectrometry, high performance liquid chromatography) of spirit drinks, beer, and traditional fermented beverages, and the valorizing of fermentation industry by-products to obtain value-added products such as functionalized beverages. Her published papers consist of the application of advanced methods for the study of volatile compounds, phenolic compounds, and risk chemicals (organochlorine pesticides, methanol, furfural, and heavy metals content) present in beverages.

Contents

Preface

Nowadays, there is a difference between food technology and food engineering. The first treats the application of methods of food preservation and food processing, while food engineering is a more complex area focusing on a combination of food and applied sciences such as microbiology, physical sciences, chemistry, and engineering.

Achieving food quality with respect to its flavor and stability is a complex process, especially when considering the large variety of flavor compounds and diversity of food accepted by consumers. Knowledge of all mechanisms involved in the change of food flavor during processing and storage and identifying the techniques for its constant quality assurance is essential. The desire for food producers to succeed in the global food market involves lengthy periods until the product reaches the consumer. Innovative cost-effective technologies are being introduced capable of satisfying consumers' quality demands.

Advanced processing methods tend to preserve the characteristic properties of food better, including its sensory and nutritional qualities, when compared with conventional food processing methods. Additionally, there is a clear rise in those suffering from food allergies. This fact is widely attributed to the changing livelihoods of populations in both developed and developing nations. The return to ancient food recipes by the industrial reinterpretation of food could be a successful alternative to a healthier lifestyle and a rising consumer trust in industrialized food.

In this book a selection of novel technologies applied to food preservation, sugar replacement, and food fortification with various bioactive compounds is presented in a series of original research and review chapters. Given the rapid growth of engineering fields, namely the food industry with novel food process technologies, novel ingredients, advanced enzyme production and applications, and other complementary technologies, this book will disclose the latest trends in food engineering. This text is a compilation of selected research articles and reviews covering current efforts in research in and application of emerging technologies in the food industry. The chapters in this book are divided into three broad sections. Section 1 deals with introductory information about enzyme application, preserving treatments (such as thermal treatment, active packaging concepts) in a sustainable, cost-effective manner, inclusion in food processing of wild edible plants as a part of cultural and generic heritage, and the upscaling of extraction techniques to increase the bioavailability of bioactive compounds. Section 2 provides data concerning the food industry's emerging technologies. Section 3 reveals the latest trends in food fortification. Overall, this book serves as an inspiring source for both scientific and industrial actors or anyone involved in any aspect related to the food industry.

I am most appreciative to Ms. Romina Skomersic, the Author Service Manager, who supervised and organized the publishing of all materials, assisted me and the authors in the completion of our work in an easy, timely manner, and provided helpful advice and guidance throughout this project. I thank the authors for their wonderful contributions. I also thank the technical editors who prepared these manuscripts for publication by IntechOpen Access. I thank too my family and colleagues, support staff, and administrators at the University of Agricultural Sciences and Veterinary Medicine of Cluj-Napoca for their support throughout this process.

Teodora Emilia Coldea, PhD
University of Agricultural Sciences and Veterinary Medicine,
Cluj-Napoca, Romania

Section 1

Introductory Chapter

Chapter 1

Introductory Chapter: A General Overview on Latest Trends in Food Engineering

Teodora Emilia Coldea

1. Introduction

Foods represent a complex mixture of hundreds of components, which contribute to their characteristic flavor. Nowadays, assuring the food's flavor stability represents an issue of food industry involved actors, becoming one of the main technological challenges. An educated consumer has increasing demands related to food quality perception over the last years. As a consequence, the food flavor instability is a critical quality issue of food industry. Achieving food's quality in respect to its flavor and stability is a complex process, especially by considering the large variety of flavor compounds and diversity of food accepted by consumers. The knowledge of all mechanisms involved in food flavor changes during processing and storage period is compulsory. Ideally, food should have the same flavor from the moment of packaging to the moment it gets to the consumer. Unfortunately, food is exposed to chemical reactions starting with the moment it is taken out from the plant. Given the actual trend of big food industry producers to accede to the global food market, this involves longer durations on which the product passes until it arrives to the consumers, contributing in increasing risk of its quality damage. New innovative cost-effective technologies are in discussion, capable of satisfying the actual consumer's quality demands.

Foods were thermally treated for cooking purposes since many centuries ago, firstly to modify and preserve their organoleptic and nutritional characteristics. Anthropologists agree that thermal cooking of food had a great impact on phenotypical properties, intellectual development of mankind, and later to the society, depending on economy, culture, and geography. The transfer from home cooking to industrialized processes began in the late nineteenth century, firstly aiming food preservation and later, after the Second World War, focusing food safety and quality issues [1]. Food engineering as a study discipline was first introduced in West European countries willing to offer a curriculum for actors in the food industry, including chemical engineering aspects, food microbiology, and biology. Nowadays, there is a difference between food technology and food engineering. The first treats the application of methods of food preservation and food processing, while food engineering is a more complex area focusing on a combination between food and applied sciences such as microbiology, physical sciences, chemistry, and engineering.

Thermal food processing is one of the most efficient methods of preservation. Beneficial effects include inactivation of food pathogens, natural toxins, and other unwanted changes, prolonging the shelf-life, adjusting the food digestibility, increasing the bioavailability of nutrients, and improving the functional properties. Still, thermal processing usually comes with unwanted changes in food composition such as decrease in nutrients, formation of toxic compounds, or other influencing the organoleptic properties. Continuously, heat treatment has to be improved according to the latest trends in food research, focusing on the increasing beneficial effects and counteracting the undesired effects.

Active packaging concept gains more interest due to its effectiveness in antimicrobial protection insurance, especially to pathogenic microorganisms. Still open for discussion, the subject refers to the influence of food-packaging polymers on food products after the incorporation of protective microorganisms. Actually, this is one of the biggest challenges when considering transposing these biomaterials to industrial scale use [2].

The past decade offered a great opportunity in valorizing the results in active biofilms characterized by antimicrobial and antifungal activities in order to extend the food shelf-life and to reduce the addition of food additives. As so, biologically active polysaccharides such as kefiran [3–5] or chitosan [6] proved an incredible potential to food industry by their high antimicrobial activity against pathogenic and spoilage microorganisms. The positive impact relates also to better environmental practices by avoiding the conventional packaging materials such as plastics and reducing the food waste [7, 8].

Advanced processing methods tend to preserve the characteristic properties of food including their organoleptic and nutritional qualities better when compared to the conventional food processing methods and, in some cases, are even more cost-effective. There is a clear rise in the populations suffering from food allergies, especially infants and children. Though this fact is widely attributed to the changing livelihood of population in both developed and developing nations and to the introduction of new food habits with advent of novel foods and new processing techniques, their complete role is still uncertain. As so, the return to ancient natural food recipes [9] could be an alternative for reinterpretation of food in a more safety and healthier way.

For example, cereals are traditionally processed into foods and beverages through fermentation worldwide. With lactic acid fermentation, as the most known ancient preserving method, studies found that valuable compounds are formed during the process [10]. Additionally, lactic acid fermentation enhances food safety by reducing the toxic compounds and producing antimicrobial factors, which facilitate the inhibition and elimination of food pathogens. Many health-enhancing attributes are related to the consumption of lactic acid fermented products.

Nowadays, demands for nondairy substitutes with high acceptance and functionality, given by pleasant flavor and nutrient content (vitamins, minerals, antioxidants, prebiotics, and probiotics), respectively, are increasing. Cereal-based beverages have a great potential to fulfill this demand [11, 12]. As part of the worldwide traditional cuisine and culture, these products are often transposed to industrial re-interpretation [13] as novel food products with various beneficial microorganisms (especially in the case of lactic acid fermented products) and functional properties designed for a balanced nutrition and life quality improvement. As so, through intensive research, focusing on this field and industrial production, these novel foods are taken beyond their national reputation. By applying novel trend technology, food products are designed for a balanced nutrition, easy and safe to store, prepare, and consume, and able to supply all the needed vitamins and nutrients.

The evolution of food industry offered many opportunities of using low-cost food additives and ingredients. Beverage industry is the most exposed to using caloric sweeteners [14], sucrose, high-fructose corn syrup, fruit juice concentrates, etc., which in latest studies proved their harmful health effects. Products such as soft drinks, carbonated soft drinks, fruit ades, fruit drinks, sports drinks, energy and vitamin water drinks, sweetened iced tea, cordial, squashes, and lemonade are attributed as the major sugar contributors in the USA for their consumption after each meal, as a supplementary source of calories. This intake severely contributes to obesity and obesity-related diseases in children and adults. As a result of novel food regulation, past years' research focused on the reduction of sugar in food and beverages along with avoiding the addition of other artificial sweeteners. Some prospects were made and proved the positive effects of a predicted reduction of sweeteners in food [15].

Wild edible plants as part of the cultural and genetic heritage are harvested and processed into value-added products especially in the Eastern Europe. They represent important sources of nutrients and health-promoting compounds, being consumed as remedies especially in rural and suburban areas. However, these products lack recognition as important contributors to human diet in developed areas [16]. Current research promotes the using of wild edible plants in human diet, while several strategies present their importance for social, economic, and agro-ecological development. Given the growing interest for a natural alternative of products obtained by chemical synthesis, bioactive compounds found increasing application in food industry, cosmetics, pharmaceutical industry, and even in agriculture.

Sea-buckthorn (*Hippophaë rhamnoides*), hawthorn (*Crataegus monogyna*), wild grown European blackberry (*Rubus fruticosus* L.), cornelian cherry (*Cornus mas*), blackthorn (*Prunus spinosa*), dog rose (*Rosa canina*), and bird cherry or hackberry (*Prunus padus*) are only few of the wild fruits with great potential for food industry as a valuable source of bioactive compounds, mainly phenolic compounds [17].

Novel bioactive compounds obtained by lab-scale extraction techniques such as high-pressure processing [18], pulsed electric field on berry fruits [19], grape pomace and food industry byproduct extraction [20–22], microwave-assisted extraction, ultrasonically assisted diffusion processes, supercritical fluid extraction, pressurized hot water extraction, gas-assisted extraction, and enzyme-assisted extraction were transposed to industrial application such as juice industry [18] and oilseed processing industry. Extraction-dependent factors—temperature, infusion duration, tea, and water ratio—were also tested in tea brewing in order to deliver important product information to consumers [23].

Modeling of processes through mathematics is widely used in food industry. Recently, an experimentally validated multiscale modeling for coffee extraction was developed [24].

Food loss is a major concern to all economic, environmental, and social aspects. In recent years, several modeling of waste treatments depending on food characteristics were proposed [25]. In order to assess the stability of a food product during storage period, degradation kinetics modeling was elaborated using different indicators such as the ascorbic acid content and color intensity loss [26].

Given the fast-growing engineering fields, namely the food industry with novel food processes technologies, novel ingredients, advanced enzyme production and applications, and other complementary technologies, this book will disclose the state of the latest trends in food engineering.

Acknowledgements

We would like to thank all authors for their contributions. Special thanks are addressed to Ms. Romina Skomersic, who assisted in the arrangement of the book and scheduling our activities.

Conflict of interest

Author declares there is no conflict of interest.

Author details

Teodora Emilia Coldea
University of Agricultural Sciences and Veterinary Medicine of Cluj-Napoca, Romania

*Address all correspondence to: teodora.coldea@usamvcluj.ro

References

[1] Van Boekel M, Fogliano V, Pellegrini N, Stanton C, Scholz G, Lalljie S, et al. A review on the beneficial aspects of food processing. Molecular Nutrition & Food Research. 2010;**54**(9):1215-1247. DOI: 10.1002/mnfr.200900608

[2] Bastarrachea L, Dhawan S, Sablan SS. Engineering properties of polymeric-based antimicrobial films for food packaging. Food Engineering Reviews. 2011;**3**:79-93. DOI: 10.1007/s12393-011-9034-8

[3] Pop C, Apostu S, Rotar AM, Semeniuc CA, Sindic M, Mabon N. FTIR spectroscopic characterization of a new biofilm obtained from kefiran. Journal of Agroalimentary Processes and Technologies. 2013;**19**:157-159

[4] Pop C, Apostu S, Salanţă L, Rotar AM, Sindic M, Mabon N, et al. Influence of different growth conditions on the kefir grains production, used in the kefiran synthesis. Bulletin UASVM Food Science and Technology. 2014;**71**:147-153. DOI: 10.15835/buasvmcn-fst:10802

[5] Pop CR, Salanţă L, Rotar AM, Semeniuc CA, Socaciu C, Sindic M. Influence of extraction conditions on characteristics of microbial polysaccharide kefiran isolated from kefir grains biomass. Journal of Food and Nutrition Research. 2016;**55**:121-130

[6] Hamed I, Özogul F, Regenstein JM. Industrial applications of crustacean by-products (Chitin, Chitosan, and Chitooligosaccharides): A review. Trends in Food Science and Technology. 2016;**48**:40-50. DOI: 10.1016/j.tifs.2015.11.007

[7] Kurek M, Garofulić IE, Tranfić Bakić M, Ščetar M, Dragović Uzelac V, Galić K. Development and evaluation of a novel antioxidant and pH indicator film based on chitosan and food waste sources of antioxidants. Food Hydrocolloids. 2018;**84**:238-246. DOI: 10.1016/j.foodhyd.2018.05.050

[8] Remya S, Mohan CO, Bindu J, Sivaraman GK, Venkateshwarlu G, Ravishankar CN. Effect of chitosan based active packaging film on the keeping quality of chilled stored Barracuda fish. Journal of Food Science and Technology. 2016;**53**:685-693. DOI: 10.1007/s13197-015-2018-6

[9] Bryła P. The role of appeals to tradition in origin food marketing. A survey among Polish consumers. Appetite. 2015;**91**:302-310. DOI: 10.1016/j.appet.2015.04.056

[10] Dueñas M, Fernández D, Hernández T, Estrella I, Muñoz R. Bioactive phenolic compounds of cowpeas (*Vigna sinensis* L). Modifications by fermentation with natural microflora and with *Lactobacillus plantarum* ATCC 14917. Journal of the Science of Food and Agriculture. 2005;**85**:297-304. DOI: 10.1002/jsfa.1924

[11] Blandino A, Al-Aseeri ME, Pandiella SS, Cantero D, Webb C. Cereal-based fermented foods and beverages. Foodservice Research International. 2003;**36**:527-543. DOI: 10.1016/S0963-9969(03)00009-7

[12] Pasqualone A, Summo C, Laddomada B, Mudura E, Coldea TE. Effect of processing variables on the physico-chemical characteristics and aroma of borş, a traditional beverage derived from wheat bran. Food Chemistry. 2018;**265**:242-252. DOI: 10.1016/j.foodchem.2018.05.095

[13] Chen KI, Erh MH, Su NW, Liu WH, Chou CC, Cheng KC. Soyfoods and soybean products: From traditional use to modern applications. Applied Microbiology and Biotechnology. 2012;**96**:9-22. DOI: 10.1007/s00253-012-4330-7

[14] Sylvetsky AC, Rother KI. Trends in the consumption of low-calorie sweeteners. Physiology & Behavior. 2016;**164**:446-450. DOI: 10.1016/j.physbeh.2016.03.030

[15] Ma Y, He FJ, Yin Y, Hashem KM, MacGregor GA. Gradual reduction of sugar in soft drinks without substitution as a strategy to reduce overweight, obesity, and type 2 diabetes: A modelling study. The Lancet. 2016;**4**:105-114. DOI: 10.1016/S2213-8587(15)00477-5

[16] Pinela J, Carvalho AM, Ferreira ICFR. Wild edible plants: Nutritional and toxicological characteristics, retrieval strategies and importance for today's society. Food and Chemical Toxicology. 2017;**110**:165-188. DOI: 10.1016/j.fct.2017.10.020

[17] Cosmulescu S, Trandafir I, Nour V. Phenolic acids and flavonoids profiles of extracts from edible wild fruits and their antioxidant properties. International Journal of Food Properties. 2017;**20**:3124-3134. DOI: 10.1080/10942912.2016.1274906

[18] Huang HW, Wu SJ, Lu JK, Shyu YT, Wang CY. Current status and future trends of high-pressure processing in food industry. Food Control. 2017;**72**:1-8. DOI: 10.1016/j.foodcont.2016.07.019

[19] Bobinaitė R, Pataro G, Lamanauskas N, Šatkauskas S, Viškelis P, Ferrari G. Application of pulsed electric field in the production of juice and extraction of bioactive compounds from blueberry fruits and their by-products. Journal of Food Science and Technology. 2015;**52**(9):5898-5905. DOI: 10.1007/s13197-014-1668-0

[20] Leong SY, Burritt DJ, Oey I. Evaluation of the anthocyanin release and health-promoting properties of Pinot Noir grape juices after pulsed electric fields. Food Chemistry. 2016;**196**:833-841. DOI: 10.1016/j.foodchem.2015.10.025

[21] Yang N, Huang K, Lyu C, Wang J. Pulsed electric field technology in the manufacturing processes of wine, beer, and rice wine: A review. Food Control. 2016;**61**:28-38. DOI: 10.1016/j.foodcont.2015.09.022

[22] Rahmanian N, MahdiJafari S, Wani TA. Bioactive profile, dehydration, extraction and application of the bioactive components of olive leaves. Trends in Food Science and Technology. 2015;**42**:150-172. DOI: 10.1016/j.tifs.2014.12.009

[23] Zhang H, Li Y, Lv Y, Jiang Y, Pan J, Duan Y, et al. Influence of brewing conditions on taste components in Fuding white tea infusions. Journal of the Institute of Brewing. 2017;**97**:2826-2833. DOI: 10.1002/jsfa.8111

[24] Moroney KM, Lee WT, O'Brien SBG, Suijver F, Marra J. Modelling of coffee extraction during brewing using multiscale methods: An experimentally validated model. Chemical Engineering Science. 2015;**137**:216-234. DOI: 10.1016/j.ces.2015.06.003

[25] Corrado S, Ardente F, Sala S, Saouter E. Modelling of food loss within life cycle assessment: From current practice towards a systematization. Journal of Cleaner Production. 2017;**140**:847-859. DOI: 10.1016/j.jclepro.2016.06.050

[26] Remini H, Mertz C, Belbahi A, Achir N, Dornier M, Madani K. Degradation kinetic modelling of ascorbic acid and colour intensity in pasteurised blood orange juice during storage. Food Chemistry. 2015;**173**:665-673. DOI: 10.1016/j.foodchem.2014.10.069

Section 2

Food Industry Emerging Technologies

Chapter 2

Correlation between Enzymatic and Non-Enzymatic Antioxidants in Several Edible Mushrooms Species

Cristiana Radulescu, Lavinia Claudia Buruleanu, Andreea Antonia Georgescu and Ioana Daniela Dulama

Abstract

Characterization of several wild growing and cultivated mushrooms from geographical area of Dambovita County, Romania, in terms of enzymatic and non-enzymatic antioxidants, through a chemometrics approach, was the aim of this study. Related to the authors' previous studies, the novelty of this paper consists in deepening research toward the complete characterization of the regional mushroom species through emphasizing their potential as food resources. In the context in which species showed their content in biological active compounds, future practical applications in the area of functional food will be developed by integrating the data concerning their lack of the toxicity and nutritional value too. Lack of data focused on the characterization of mushroom species investigated in the paper supports the significance of this research. The statistical analysis of data highlights the relationship between compounds showing antioxidant activity of autochthonous mushrooms (both cap and stipe).

Keywords: mushrooms, antioxidant activity, polyphenols, flavonoids, enzymes

1. Morphological characteristics and nutritional values of wild and cultivated autochthonous edible mushrooms

The mushrooms contain an extremely wide variety (over 100,000 species spread across all ecosystems) and are traditionally included among the plants. Currently, they are considered as a self-standing group, halfway between the plant world and the animal world. It is true that due to their anatomical and physiological characteristics, they present traits of both worlds (i.e. plant and animal). They are distinguished from chlorophyll plants by the total absence of photosensitizing pigments, hence the inability to produce sugars and starch, starting from the carbonic anhydride present in the atmosphere [1–4]. The mushrooms have their cell walls made from cellulose, but also from chitin, an insect-specific component. So, to survive, mushrooms consume simple substances (such as proteins and sugars) produced by others [2].

It is well known that macromycetes (i.e. superior fungi) represent a heterogeneous group that includes both *Ascomycota* and *Basidiomycota* phyla, with a fruiting body of at least 4 mm in diameter [2]. In Romania [2–4] are known over 2500 varieties of basidiomycetes, of which more than 500 are edible, being more or less tasty. During their development, superior fungus passes through two successive stages [3, 4]: (a) vegetative, represented by the mycelium from substrate, the body or talus of the fungus; (b) fertile, in which, the fruiting body (i.e. spore producer) is formed. From a morphological point of view, the fruiting body of mushrooms is consisted by pileus (i.e. cap) and stipe (i.e. stalk, stem). Consequently, basidiomycetes have a typical umbrella, with one stipe more or less developed, either cylindrical or thickened or subdued at the base, sometimes flattened or extended in soil with a variable length mycelium. It is important to observe how to insert the stipe into the cap: it can be completely central, more or less eccentric, and sometimes completely lateral [2, 3]. During the fungus growth, the pileus form varies widely: at first it is shaped globes and at complete maturation may be concave or funnel-shaped [2, 3].

The vegetative organ of mushroom consists from a hyphae mycelium that is gathered and placed in a soft fiber texture (real mycelium). The macromycetes mycelium can live outside of the organism on which of it develops (epiphytic) or inside of it (endophytic). Epiphytic fungi sometimes lead a saprophytic life and the endophytes are always parasitic. Nutrition of macromycetes is heterotrophic, the mycelium acting on the nutrient substrate, live or non-living, through the enzymes which it secretes [2, 3].

The compact bodies of mushrooms appear to the substrate surface, either solitary, either in irregularly arranged groups, linearly or in circles, either in the form of bushes [3–5]. Favorable conditions of nutrition, temperature and humidity, are the main factors involved in superior fungi growing [3–9].

The most common basidiomycetes families, which are growing in the Romanian forests, appreciated from nutritional point of view, are: *Lepiotaceae* (e.g. *Macrolepiota mastoidea*, *Macrolepiota rhacodes*), *Tricholomataceae* (e.g. *Tricholoma rutilans*, *Tricholoma columbetta*, *Tricholoma terreum*), *Hygrophoraceae* (e.g. *Hygrophorus marzuolus*, *Hygrophorus eburneus*), *Russulaceae* (e.g. *Russula aeruginea*, *Russula alutacea*, *Russula atropurpurea*, *Russula cyanoxantha*, *Russula delica*, *Russula nigricans*, *Russula vesca*), *Pleurotaceae* (e.g. *Pleurotus cornucopiae*, *Pleurotus ostreatus*), *Agaricaceae* (e.g. *Agaricus augustus*, *Agaricus campestris*, *Agaricus bisporus*, *Agaricus silvaticus*, *Macrolepiota procera*, *Macrolepiota excoriata*), *Boletaceae* (e.g. *Boletus edulis*, *Boletus pinicola*, *Boletus aereus*, *Boletus regius*), *Fistulinaceae* (e.g. *Fistulina hepatica*), *Cortinariaceae* (e.g. *Cortinarius varius*, *Cortinarius caperatus*, *Cortinarius collinitus*), *Amanitaceae* (e.g. *Amanita rubescens*, *Amanita citrina*), *Cantarellaceae* (e.g. *Cantharellus cibarius*, *Cantharellus lutescens*), *Physalacriaceae* (e.g. *Armillaria mellea*) and so on.

Particularly, in the forests of Dambovita County several species such as, *Russula atropurpurea*, *Russula cyanoxantha*, *Russula alutacea*, *Russula nigricans*, *Russula vesca*, *Pleurotus ostreatus*, *Armillaria mellea*, *Cantharellus cibarius*, *Boletus edulis*, *Macrolepiota excoriate*, *Macrolepiota procera*, *Agaricus bisporus*, *Agaricus campestris* are very widespread (**Table 1**). From this reason, these species were characterized from morphological and nutritional point of view (**Table 2**).

Mushrooms provide several important nutrients (**Table 2**) which reduced risk of obesity [5] and overall mortality [6], diabetes [7], and heart disease [8]. In this respect, mushrooms are a high content in antioxidants [9], selenium [21], vitamin D [22] and folic acid [23], substances which inhibit the growth of cancer cells [19] by contributing to the regulation of the cell growth cycle [24]. Dietary fibers (i.e. beta-glucan and chitin) from superior fungi also benefit the digestive system [1] and reduce the risk of heart disease and metabolic syndrome [25]. Together with fibers, the high content of potassium and vitamin C in mushrooms decrease the risk of high blood pressure and cardiovascular diseases [26, 27]. Several other minerals

Aspect	Scientific classification: division/class/order/family/genus/specie	Habitat
	Basidiomycota/Agaricomycetes/Russulales/Russulaceae/*Russula*/*Russula atropurpurea*	Conifer forests, under broad-leaf trees of oak and beech [3]
	Basidiomycota/Agaricomycetes/Russulales/Russulaceae/*Russula*/*Russula alutacea*	Hornbeam areas [3]
	Basidiomycota/Agaricomycetes/Russulales/Russulaceae/*Russula*/*Russula nigricans*	Durmast forest [3]
	Basidiomycota/Agaricomycetes/Russulales/Russulaceae/*Russula*/*Russula cyanoxantha*	Durmast and hornbeam areas [3]
	Basidiomycota/Agaricomycetes/Russulales/Russulaceae/*Russula*/*Russula vesca*	Deciduous forests [3]

Aspect	Scientific classification: division/class/order/family/genus/specie	Habitat
	Basidiomycota/Agaricomycetes/Agaricales/Pleurotaceae/*Pleurotus*/*Pleurotus ostreatus*	On trunks of deciduous species [3]
	Basidiomycota/Agaricomycetes/Agaricales/Physalacriaceae/*Armillaria*/*Armillaria mellea*	Grows solitary or in groups, on trunks of oak and beech but also on conifer trunks, roots, rotten logs [3, 4]
	Basidiomycota/Agaricomycetes/Cantharellales/Cantharellaceae/*Cantharellus*/*Cantharellus cibarius*	Beech and conifer forests [3]
	Basidiomycota/Agaricomycetes/Boletales/Boletaceae/*Boletus*/*Boletus edulis*	Conifer forests [3]
	Basidiomycota/Agaricomycetes/Agaricales/Agaricaceae/*Macrolepiota*/*Macrolepiota procera*	Open woods and pastures as well as besides the paths in the forests (e.g. oak and beech or coniferous) [3]

Aspect	Scientific classification: division/class/order/family/genus/specie	Habitat
	Basidiomycota/Agaricomycetes/Agaricales/Agaricaceae/*Macrolepiota*/*Macrolepiota excoriata*	Pasture, heaths and open woodland [3]
	Basidiomycota/Agaricomycetes/Agaricales/Agaricaceae/*Agaricus*/*Agaricus bisporus*	Grassy areas following rain or forest edge [3]
	Basidiomycota/Agaricomycetes/Agaricales/Agaricaceae/*Agaricus*/*Agaricus campestris*	Meadows, wet grasslands or forest edge [3]

Table 1.
Scientific classification and habitat of several autochthonous edible mushroom species.

copper, iron, and phosphorus, are available in mushrooms and according with their species, these elements and more (i.e. zinc, manganese, sodium, calcium) can be accumulated from their habitat and translocated in stipe and cap in different concentration [26, 28]. Sometimes high concentration in heavy metals such as, lead, nickel, cadmium, chromium was found in mushrooms species [4, 29–31].

Some species, mainly from genera *Agaricus*, *Macrolepiota*, and *Russulaceae*, accumulate high levels of cadmium and lead even in unpolluted and mildly polluted areas [32]. The concentrations of both metals and also of chromium and nickel increase considerably in the heavily polluted sites, such as in the vicinity of metal smelters [30]. Present knowledge of metal speciation in mushrooms is limited as well as is knowledge of their bioavailability in human's body. Thus, consumption of

Mushrooms	Morphological characteristics	Nutritional value
Russula atropurpurea	Convex/flat cap (4–10 cm in diameter) and color dark purple and often almost black in the center; loud stipe (length 3–6 cm, diameter 1–2 cm); whitish spores ornamented with warts and ridges, 7–9 × 6–7 μm measure [3]	Vitamins C, D, B, choline, folic acid, chitin and beta-glucans, potassium, phosphorous, iron, copper [4]
Russula alutacea	Compact and fleshy hemispheric cap then flat at maturity; 7–13 cm diameter cap, purplish to black color; white and dense internal tissue, with low flavor of fruit and sweet taste [3]	Fe (118.17–130.88 mg/kg d.w.); Cu (13.28–14.19 mg/kg d.w.); Zn (15.11–17.84 mg/kg d.w.); Mn (19.58–26.76 mg/kg d.w.) [10]
Russula nigricans	Hemispherical cap, first white, with a dent in the center, then brown, flattened at the end of the maturation phenophase, when it becomes deep, in the cup form, with the edge at first begging to the foot, then high and wavy; consistent and short stipe, cylindrical or thin at the base [3]	Fe (107.03–141.30 mg/kg d.w.); Cu (6.72–13.10 mg/kg d.w.); Zn (25.41–94.81 mg/kg d.w.); Mn (34.21–57.41 mg/kg d.w.) [10]
Russula cyanoxantha	Compact and fleshy cap initial globular, then flat and deep on center, concave, with edge raked to stipe then stretched or wavy; blunt and fleshy stipe, thickened at middle and thin on the base, with rough and flour-like surface [3]	Carbohydrate (9.56%); protein (49.2%), fat (7.87%), crude fiber (30.81%), ash (2.56%) [11]
Russula vesca	Cap is 5–10 cm wide, flat, convex, or with slightly depressed centre, weakly sticky, color brownish to dark brick-red; stipe narrows toward the base, 2–7 cm long, 1.5–2.5 cm wide, white; taste mild. White spore print [3]	Carbohydrates (70.9%); crude protein (25.71 g/100 g d.w.); lipid (3.07 g/100 g d.w.); crude fiber (5.18 g/100 g d.w.); ash (6.82 g/100 g d.w.); magnesium (14 g/kg d.w.); calcium (31 g/kg d.w.); potassium (2.2 g/kg d.w.) [12]
Pleurotus ostreatus	Very fleshy cap with 5–15 cm diameter, convex at the beginning and then flat and at maturing time is thickens and deepens, in the seashell form, with smooth surface and glossy, often wavy; white and compact internal tissue with pleasant smell [3]	Protein (26.67%); ash (9.83%); crude fiber (11.05%); potassium (22.81 mg/100 g d.w.) [13]
Armillaria mellea	Yellow and fairly consistent cap with a diameter of 4–15 cm, first hemispherical, then flat, and at the end of the maturity phenophase slightly deeper; smooth, glossy cuticle on the wet weather and matte on dry time; 5–20 cm tall of stipe, cylindrical, brown, elongated, bulbous base, fluffy, tough and elastic consistency; white spores [3]	Well-balanced nutrients: δ-tocopherol 42.41 μg/100 g d.w.; carbohydrates (81.25 g/100 g d.w.), ash (8.84 g/100 g d.w.), fat (1.97 g/100 g d.w.), proteins (1.81 g/100 g d.w.); polyunsaturated, saturated, monounsaturated fatty acids [14]
Cantharellus cibarius	Compact, hard, yellow and fleshy cap, with diameter of 3–10 cm, convex at first, then slightly deep, and at full maturity it takes the shape of a deeply funnel, with irregular surface and corrugated edge; yellow, robust, hard, smooth stipe, 3–8 cm height, in the frustum cone shape; spores are yellow [3]	Vitamin C (0.4 mg/g fresh weight), potassium (~0.5%, fresh weight), vitamin D, ergocalciferol (vitamin D2 212 IU/100 grams fresh weight) [1]; 20.21% saturated acids, 77.69% unsaturated fatty acid, 17.92% monosaturated acids, 59.79% polysaturated acids, 12.81% palmitic acid, 59.79% dien, 13.57% oleic acid, 59.79% linolenic acid [15]

Mushrooms	Morphological characteristics	Nutritional value
Boletus edulis	Cap diameter is 5–25 cm, brown, very fleshy and hard, initially hemispherical, then convex; compact and solid stipe, cylindrical, thickened or even globular, with a very fine and dense surface; brown spores [3]	Carbohydrates (65.4% d.w.) [16]; palmitic acid, 9.8%; stearic acid, 2.7%; oleic acid, 36.1%; linoleic acid, 42.2%, linolenic acid, 0.2% [17]; 20 essential and nonessential amino acids, (total content 2.3 g/100 g d.w.) vitamin D2 (4.7 μg/100 g d.w.); selenium (13–17 ppm) [18]
Macrolepiota procera	Cap has 10–30 cm in diameter, initially globular, then convex, umbrella-shaped, with a dark-brown, smooth central gurgle; brown, cylindrical stipe, with 10–40 cm height, is hollow inside, long, compact and fragile, bulbous at the base, provided with a large, membranous and strong ring, can slide down along stipe [3]	Carbohydrates (glycerol, mannitol, glucose, trehalose, lepiotan); 15.9% saturated acids, 81.95% unsaturated fatty acid, 19.51% monosaturated acids, 62.44% polysaturated acids, 10.95% palmitic acid, 62.44% dien, 17.40% oleic acid, 62.44% linoleic acid; chitin, proteins, fiber, vitamins, minerals [3, 15]
Macrolepiota excoriata	First ovoid, then flat, with a large gurgle in the middle, peanut shell color, in a darker shade in the center; cylindrical, fusiform stipe, bulbous at the base, whitish above the ring and straw-colored beneath it with flour-like aspect [3]	68.4% carbohydrates, 23.9% crude protein, 5.4% ash, 68.59% saturated acids, 26.58% unsaturated fatty acid, 26.58% monosaturated acids, 45.06% palmitic acid, 8.81% oleic acid [15, 16]
Agaricus bisporus	Cap is a pale gray-brown in color, with broad, flat scales on a paler background and fading toward the margins; first is hemispherical in shape before flattening out with maturity, and 5–10 cm in diameter; cylindrical stipe is up to 6 cm tall by 1–2 cm wide and bears a thick and narrow ring, which may be streaked on the upper side; spore print is dark brown, oval to round, measuring 4.5–5.5 μm × 5–7.5 μm [2, 3]	Carbohydrates (3.26 g/100 g d.w.); vitamins, such as: 7% thiamine (B1), 34% riboflavin (B2), 24% niacin (B3) 30% pantothenic acid (B5), 8% vitamin B6, 4% folic acid (B9), 2% vitamin B12, 3% vitamin C, 1% vitamin D; protein (3.09/100 g) [19]
Agaricus campestris	Diameter of cap is 5–15 cm; first, it is globular, then hemispherical and stretched, perfectly flat in center, white, silky and smooth, with brown flakiness; stipe has tall of 3–7 cm and thick of 1–3 cm, slightly narrowed to the base, white, full, hard, smooth, squamous under the ring; small and fragile ring; pseudo-tissue is soft inside, white, with a pleasant smell and taste	Carbohydrates (30.4%), proteins (18.9%), polyunsaturated acid (34.4%), monounsaturated acid (48.4%), saturated acids (17.2%), potassium (66.5%), phosphorous (21.2%) [20]

Table 2.
Morphological and nutritional characteristics of several autochthonous edible mushroom species.

the accumulating species should be restricted. The cultivated species, especially the oyster mushroom (*Pleurotus ostreatus*) contain only low levels of the trace elements according with previous authors studies [4, 30].

2. Mushrooms as functional foods

The main role of the diet is to provide adequate nutrients in satisfactory quantities for the metabolic needs of the body and, in addition, to give to consumer a sense of satisfaction and pleasure through the hedonic attributes of food. Recent research supports the hypothesis that, besides to meeting nutritional needs, diet can modulate various physiological functions and may play unfavorable or beneficial roles in some diseases. It is been seen recently the beginning of a new era in nutrition, reflected by changing the consumer' attitude and manifested through: awareness of the connection between physical and mental status, respectively food, as well as between diet, longevity and physical appearance; attention paid to health promoting compounds (antioxidants, vitamins, etc.); the belief that the diet can provide more promising health solutions than the medical cabinet.

Foods designed to improve people's health and for which claims on specific health effects are allowed were introduced on the market at the beginning of the functional foods era as Foods for Specified Health Use—FOSHU, specific criteria for their labeling being defined. According to European Commission [33], many definitions of functional foods are met worldwide, without being official or commonly accepted. A definition proposed by European Commission Concerted Action on Functional Food Science in Europe—FUFOSE (Consensus document on "Scientific Concepts of Functional Foods in Europe") [34] for functional foods is the following: *"food that beneficially affects one or more target functions in the body beyond adequate nutritional effects in a way that is relevant to either an improved state of health and well-being and/or reduction of risk of disease"*. Costin and Segal [35], defined functional foods as *"food products and their compounds that improve the general health of the consumers, avoid the risk of illness, improve the physical or psychological quality of life, as well as the recovery capacity after extenuating exercise or various illnesses"*.

There are more terms linking food and nutrition with health, such as *food supplements/dietary supplements*, *nutraceuticals/nutriceuticals*, *pharmafood*, *designer food*. Clear definition and consistent legislation are challenges for all parties (policy makers, producers, researchers and so on) involved in issues related to functional foods. Scientific substantiation of health claims through integration of different disciplines requires clinical studies to prove the effectiveness of functional foods on large-scale. The credibility of health claims is considered a condition for the success of functional foods in economically terms.

Mushrooms are well known for their nutritional value and health-promoting properties, being considered as both functional foods and a source of nutraceuticals [36]. This is due to biologically and physiologically active substances such as phenolic acids [37]. Recently, a growing interest related to mushroom's mechanisms of action was observed. Thus, it seemed that such mushrooms species can prevent diseases correlated with increased formation of free radicals and oxidative stress, due to their antioxidant capacity. It is reported that another bioactive property, the antimicrobial activity of mushroom extracts, could have a positive influence in evolution of chronic diseases (diabetes, cardiovascular diseases or various types of cancer) [38]. According with Kalaras et al. [39], mushroom consumption may be associated with reductions in oxidative stress-related diseases and disorder because mushrooms species (particularly the yellow oyster and porcini) are rich in glutathione (GSH) and sulfur-containing amino acid ergothioneine (ERGO), considered critical antioxidants. Part of them was determined as uniquely high in both GSH and ERGO. Not in the last, the antimicrobial activity of the mushrooms extracts and their phenolic acids was concretized in strong antibacterial and antifungal properties [40]. In several cases these ones were reported as higher than those of the antibiotics and antifungals frequently used. Some mushrooms species revealed demelanizing properties, in different proportions against different fungi. Their phenolic extracts showed highest demelanizing abilities [41].

In the past years, an increasing public awareness of potential health benefits of dietary fibers was observed. Thus, the food producers tried to fulfill the consumers' request by developing a wide range of fiber-enriched or fiber-fortified foods. In this context, edible mushrooms are taken into account as a rich source of some novel dietary fibers (DF), with beneficial health effects to humans [42]. Health benefits associated to DF from mushrooms include blood glucose and lipid attenuation, antitumor activity as well as immune-enhancing. Immunomodulating and antitumor effects of mushrooms and their extracts are attributed primarily to content in beta-glucans or polysaccharide-protein complexes [8]. Part of health benefits are supported on the basis of *in vitro* and *in vivo* animal trials. Stronger health benefits/effects were reported for mushroom components/extracts than whole mushrooms in a small number of direct human trials.

The mushrooms can be considered an important and valuable resource for practical applications in the area of functional food, not only as dietary food. An overview on this topic [43] pointed out that *Agaricus bisporus* combined with dried dates for producing white bread lead to the improvement of protein, iron and other nutrients in quantitative and qualitative terms. Addition of powder obtained from the same mushroom species to obtain sponge cakes exhibited acceptable sensory characteristics and a better nutritional value. An extruded cereal-based product obtained with *Cordyceps militaris* proved to have significant anti-fatigue property compared to the product of cereal grains.

Designing new functional foods from mushrooms supposes to preserve their biologically active compounds. Water or organic solvents used for conventional extractions could lead to noticeable degradation of these components. Because consumers asked “natural”, “safe” and with “nutritional value” foods from plant materials, processed with sustainable methods [44], obtaining extracts through techniques aligned with the “green” concepts, such as microwave-assisted extraction, high-pressure assisted extraction, pulsed electric fields assisted extraction or ultrasound-assisted extraction [45] are preferred. These novel non-conventional methods, including also subcritical and supercritical fluid extraction or enzyme-assisted extraction for recovery of valuable compounds from mushrooms, are environmentally friendly methods for production of nutraceuticals or various ingredients for functional foods [46].

Corrêa et al. [47] proposed the production of a natural extract rich in ergosterol as added-value food ingredient, by using a commercially discarded *Agaricus blazei* fruiting bodies. With a significant antioxidant and antimicrobial properties and showing no hepatotoxicity, this extract was used as fortifier ingredient for yogurts. According to authors, the circular bioeconomy concept is fulfilled too, having in view that *A. blazei* fruiting bodies are normally discarded, being inconsistent to the commercial requirements of the market.

Dried synbiotic foods, shelf stable and economical advantageous, were formulated by Moumita et al. [48], using *Enterococcus faecium* as probiotic and *Pleurotus florida* extract as prebiotic. Lyophilization and spray-drying lead to microcapsules added to different dry food matrices. Choosing to *Pleurotus florida* was due to its content in β-glucan, a well-known prebiotic which stimulates selectively the growth of probiotics.

The prebiotic potential of polysaccharides from different mushroom species was in focus for other researchers [49]. Thus, studying 53 wild-growing mushrooms, the authors found that the majority of their polysaccharides stimulated the growth of *Lactobacillus acidophilus* and *Lactobacillus rhamnosus* isolated from the human gastrointestinal tract. For this reason, the polysaccharides fractions from edible mushrooms could enhance the number of beneficial bacteria in the GI tract, being useful in producing functional foods and nutraceuticals. Microencapsulation of alcoholic extracts of *Agaricus bisporus* was made by spray-drying, using maltodextrin cross-linked with citric acid [50]. The microspheres with the extracts protected this way by degradation were used to obtain functionalized yogurts, with promising bioactive properties.

Functional food products with better characteristics in terms of stability were obtained through addition of mushroom powder to meat emulsion (batters) [51].

Mushrooms (*Agaricus bisporus*) were taken into account having in view by one hand their availability as plant protein source and by another hand the consumers' interest in a balanced ratio of plant/animal protein intake. Meat emulsions with 2% mushroom powder added were proved to lead to a well-ordered emulsion structure, due to their higher protein adsorption at the lipid interface. These ones exhibited heat resistance and higher gel-like behavior compared with other samples. Obtaining final products through cooking the meat emulsions, an improvement in textural properties was determined. Considering color as one of the most important properties from the consumers' viewpoint, so that they would like to purchase the final products of meat emulsions, the authors established, in the model meat emulsion (without addition of nitrite curing salt, spice mix or additive mix), that increasing the mushroom powder lead to increasing of the redness values of the emulsion and the decreasing of their lightness and yellowness.

The protein and soluble and insoluble dietary fiber content of mushroom was by interest for other researchers too. Thus, the mushroom powder was added to pasta, considered a nutritional imbalanced food [52]. This one lead not only to the deficiency made up, to decreasing of the extent of starch degradation, but also to production of food with health-promoting bioactive compounds. Thus, the mushroom powder was proved to confer healthier characteristics, such as improving antioxidant capacity and lowering the potential glycemic response when are incorporated into fresh semolina pasta.

A β-glucan composed mainly of polysaccharide with some proteins and a small amount of phenolic compounds, extracted from *Agaricus bisporus*, *Pleurotus ostreatus* and *Coprinus attrimentarius* was proved to have antioxidant activities and functional properties [53]. In terms of functional properties, the β-glucan from *C. attrimentarius* showed the highest fat binding, emulsifying properties and swelling power, while the one from *P. ostreatus* exhibited the highest foaming properties. In authors' opinion, this mushroom β-glucan could be an effective functional ingredient for food formulations and pharmaceutical ones too. The food applications of *Pleurotus* powder or β-glucan-rich fractions isolated from *Pleurotus* spp. are considered well known and described [54]. Emphasizing that the perspectives for *Pleurotus* spp. applications in functional foods are related to consumers' acceptability, the authors summarized its three main strategies of development, respectively the use as fortifying agent, as high-cost protein replacer and as prebiotic ingredient too. Consumption of functional foods containing specific extracts from mushrooms should be encouraged among people needed to lower their cholesterol levels in serum [55]. This statement is based on the *in vitro* and *in vivo* studies showing that fungal extracts obtained from edible mushrooms (β-glucans and other water-soluble compounds) might be able, as pharmaceutical drugs and functional foods, to modulate cholesterol levels.

The molecular basis underlying the biologically active compounds benefits to human health in the case of certain mushrooms species seems to be not yet elucidated. Examining the biological effects of the MeOH extract of *Morchella esculenta* L. (Morchellaceae), commonly known as morel mushroom (found throughout the world, widely distributed in Korea, China, Japan and Europe), Lee et al. [56] isolated eight compounds (three fatty acids and five sterols) that exhibited potent cytotoxicity to human lung cancer cell lines. In the authors' opinion, further evaluation would provide the evidence for the use of *M. esculenta* as functional food against cancer, with significant implications for cancer—the second leading cause of mortality worldwide—prevention and treatment. *Grifola frondosa* (known as maitake mushroom), used widely as a daily food, as food additive or for medicinal reasons, was investigated by Dissanayake et al. [57] with a view to evaluate its functional food value. The highly abundant phytochemicals determined (glycerides, sterols, a glucosylceramide, a α-glucose dimer, a phospholipid and α-glucans) seems to be responsible for the antiinflammatory and antioxidant activities of the fruiting body of *G. frondosa*. The authors concluded that health benefits and an improved quality of life could be achieved through a regular consumption of maitake mushroom.

In order to provide the scientific evidence needed for the development of functional food for the management of certain health problems afflicting millions of people worldwide, the study of Akata et al. [58] showed that *Lycoperdon utriforme* and *Agaricus campestris*, due to their biologically active constituents, can be used for prevention of diabetes type II and Alzheimer's disease. Mushrooms that have demonstrated experimental or/and clinical anti-diabetic effectiveness are also *Tremella fuciformis* (berk), *Wolfiporia extensa* (Peck) Ginns, *Ganoderma lucidum* (Curtis) P. Karst, *Ganoderma applanatum* (Pers.) Pat., *Collybia confluens* (Pers.: Fr.) Kummer, *Auricularia auricula-judae* (Bull.), *Agaricus subrufescens* (Peck), *Inonotus obliquus* (L.), *Hericium erinaceus* (Bull.), *Agrocybe aegerita*, *Coprinus comatus* (O.F. Mull), *Cordyceps sinensis* and *Grifola frondosa* (Dicks.) [59].

Limited knowledge about the processing effects on the mushrooms biologically active compounds or the mushrooms and their derived products functional properties still persists, despite the advances in research from the last decades. Moreover, various interactions within the food matrices if mushrooms or their ingredients are added, are yet unknown.

Nutrition represents a psycho-social act, because it cannot be understood only as a satisfaction of certain nutritional needs. Food means nourishment, stimulus for emotional tonus and a symbolic significance that the individual gives to food. Food is to be enjoyed and the physiologically eating is a complicated process.

For these reasons, the sensory profile of food represents an overwhelming concern of all producers, so that to fulfill the consumers' needs. In this context and on the basis of the increased interest in mushrooms in terms of their nutritional and bioactive compounds, the research on their sensory properties is considered surprisingly scarce [60]. With a view to add knowledge in this gap, the authors processed edible mushrooms (*Boletus edulis*, *Cantharellus cibarius*, *Craterellus tubaeformis*, *Lactarius camphorates* and *Agaricus bisporus*) by sous vide cooking, frozen, pooled and tempered to 50–60°C. The sensory evaluation made by a trained panel revealed a moderately intense total odor for all samples. Weak cardboard-like, forest-like and earthy odor notes were defined for the mushroom species above mentioned. In terms of taste, *Agaricus bisporus* was moderately sweet and most intensively umami-like. *Boletus edulis* closely followed the umami intensity of *A. bisporus*, being also the sweetest sample. *Lactarius camphorates* was characterized as very different from other samples, due to its sensory profile, defined as intensively bitter and astringent. This curry milk cap, used in Finland often as spice unlike other mushroom species, was the only noticeably pungent sample. Aisala et al. [60] demonstrated, on the basis of projective mapping applied to three wild and three cultivated types of mushrooms blanched, frozen and thawed to ambient temperature, the major influence of processing on the sensory properties of the mushrooms. It seems that most consumers choose to avoid eating button mushroom because the fresh samples are linked to umami and mushroom descriptors. The other species were linked to umami and mushroom descriptors. Finally it was concluded that wild and cultivated mushrooms are different in sensory descriptors and their related intensities too. Varying profiles, new innovative mushroom products and food ingredients could be designed.

In order to meet the concern related to major nutritional problems in most countries (protein energy malnutrition and micronutrient deficiencies), Ishara et al. [61] tested the use of mushroom (*Agaricus bisporus* and *Pleurotus ostreatus*) flours in different blends with maize flour, in nutritional and functional terms. The mushrooms flours could serve as protein supplements and food fortification, due to their increased protein, fibers and mineral content. The composite flours noticed through an increased water retention capacity, water absorption capacity, foaming capacity, fat absorption capacity and a decreased bulk density and syneresis. These data indicated that mushroom flours can be very suitable in human diet.

Drying methods (convective drying, freeze-drying, vacuum microwave drying and a combination of convective predrying and vacuum microwave finish drying) were tested in order to evaluate their influence on the sensory profile and implicitly on the quality of the oyster mushrooms (*Pleurotus ostreatus* Jacq.) [62]. The total concentration of aroma/volatiles compounds of fresh mushroom was reduced significantly by all drying treatments. However, the combined treatment mentioned above leads to obtaining products with a sensory profile closer to the fresh mushrooms. Nonthermal plasma technology (NTPT) was investigated in order to better understanding of the mechanism of interaction of food bioactive compounds and plasma and consequently its successful adoption by industry. Reviewing the influence of NTPT on functional food components, Muhammad et al. [63] showed that the plasma activated water (PAW) has the effect of increasing the antioxidant activity and the concentration of ascorbic acid of button mushroom. The antioxidant activity was extended with increases in PAW processing time.

3. Evaluation of several antioxidant species in indigenous edible mushrooms

As described in previous chapter, enzymatic and non-enzymatic antioxidants were reported to be present in edible mushrooms, and having roles in balancing human metabolic processes related to oxidative stress [64]. Since mid of twentieth century, after Harman's "Free-Radical Theory of Ageing", an intensive research on involvement of free radicals and antioxidants in living processes has been performed [64]. They are commonly named *reactive species* the advanced knowledge allowed their classification in terms of intensity [65], as well as of the nature of active centers (either oxygen, nitrogen, carbon or sulfur species) [64].

Wild grown and cultivated mushroom species have been studied from the perspective of possible correlations between non-enzymatic and enzymatic antioxidants. Thus, for the first category several phytochemical characteristics have been determined: total phenolic content, total flavonoids, antioxidant activity, and four trace micronutrients (Zn, Fe, Mn, and Cu), and for enzymatic antioxidants data, activity of catalase (CAT) and peroxidase (POX) have been evaluated.

3.1 Analytical techniques

Different instrumental analytical techniques were reported by authors [66] to identify and quantify antioxidants in edible mushrooms. Among these, high performance chromatography (HPLC) and gas chromatography (GC) using various detection devices, spectroscopic techniques such as nuclear magnetic resonance (RMN), Fourier transform infrared (FTIR), ultraviolet-visible (UV-VIS), as well as inductively coupled plasma mass spectrometry (ICP-MS) are among the most applied. According to available equipment, the mushroom samples were characterized mainly through the absorption spectroscopic methods UV-VIS and FTIR spectroscopy, and information on trace micro-nutrients was gathered through ICP-MS. Analytical process included the classic steps of (a) sampling, (b) sample treatment and/or preparation, and (c) qualitative and/or quantitative analysis. For the step (a), mushroom samples were collected from the natural habitat according with **Table 1** and then representative portions from each sample were taken for further treatment. For the step (b), several procedures have been used: (i) oven drying at 40°C for 48 h; (ii) grounding to less than 2 mm, cap and stipe separately; (iii) extraction of analyte(s), for 4 h, at room temperature, under continuous mixing, in two type of solvents—redistilled water and hydroalcoholic 50% (v/v), dry matter (g) to solvent ratio (mL) was of 4:100; (iv) centrifugation of obtained extracts; (v) wet digestion with concentrated HNO_3 and H_2O_2, at ratios of 0.2 g dry mushroom sample to 50 mL solution.

3.1.1 Total content of polyphenols (TCP)

Polyphenols are a class of compounds with structures containing at least one aromatic ring with at least one hydroxyl group bonded on it. They are classified according to the number of rings and to their functional groups bound in the structure, and thus we have: phenolic acids, flavonoids, stilbenes, and lignans, coumarins, tannins. Phenolic acids were reported to be the main polyphenolic compounds in mushrooms [37].

Total phenolic content in mushrooms was reported to be successfully determined by Folin Ciocalteu method [67], and an adapted method was applied for the studied mushrooms [67]. The Folin Ciocalteu reagent is a mixture prepared by dissolving sodium tungstate ($Na_2WO_4{\cdot}2H_2O$) and sodium molybdate ($Na_2MoO_4{\cdot}2H_2O$) in water with hydrochloric acid and phosphoric acid. Hydrated lithium sulfate ($Li_2SO_4{\cdot}H_2O$) may be added to this mixture to prevent turbidity that may appear due to formation of some insoluble sodium salts [67]. The mixture is very stable if protected to reduction agents and light. Diluted reagent also needs to be protected to light. The chemical process, occurring at basic pH, is based on molybdenum reduction from +6 (yellow) to +4 (blue) after the oxidation of polyphenols in samples, and may be described in **Figure 1** [68].

Light absorption of a monochromatic radiation of 765 nm is measured with a UV-VIS spectrophotometer. Colored liquid samples were placed in 10 mm light path cuvettes and readings were made versus a blank sample prepared with all reagents as samples, but with extractant instead of mushrooms extract. A calibration curve with gallic acid as reference antioxidant was plotted before each measurement set, calibration range chosen was 0.01–0.08 mg/mL.

Similar experimental procedures were applied for both aqueous and hydroalcoholic extracts, different samples dilutions were used so that the linear domain of Beer-Lambert-Bouguer law and calibration range were reached.

Total polyphenols content were expressed as milligrams of gallic acid equivalents per mL of extract, and then reported to mushroom dry weight (mg GAE/g d.w.). All experiments were performed in triplicate and the means ± standard deviations (SD) were reported [69].

3.1.2 Total flavonoid content (TFC)

Flavonoids are antioxidant compounds whose structure has two benzene rings (A and B) and an oxygen containing pyran ring (C). Six subclasses of flavonoids are generally accepted for classification, as follows: flavonols, flavones, isoflavones, flavanones, anthocyanidins and flavonols [70–72], differentiated by the oxidation

$$Na_2WO_4/Na_2MoO_4 \text{ (yellow)} \longrightarrow (\text{Fenol-}MoW_{11}O_{40})^{-4} \text{ (blue)}$$
$$Mo^{+6} + e^- \longrightarrow Mo^{+5} \text{ (blue)}$$
$$Mo^{+5} + e^- \longrightarrow Mo^{+4} \text{ (blue)}$$

Figure 1.
Reaction scheme for total polyphenols determinations by Folin Ciocalteu spectrometric procedure.

level of the C ring of the basic 4-oxoflavonoid (2-phenyl-benzo-γ-pyrone) nucleus. Presence of flavonoids in edible mushroom extracts has been confirmed by several authors [67, 69, 73], some of these reported also their molecular identification (i.e. myricetin, chrysin, catechin, resveratrol, quercetin, others). The antioxidant activity of flavonoids, as for polyphenolics in general, is mainly given by the presence and position of multiple hydroxyl groups in their molecules. Thus, it is considered that the primary mechanism of the radicals scavenging activity of flavonoids is hydrogen atom donation [70, 71].

Total flavonoid content in aqueous and hydro-alcoholic mushroom extracts was measured by the aluminum chloride colorimetric assay described in the literature [74], adapted for the working conditions [69]. Method's principle is based on Al^{3+} ions to form complex combinations with carbonyl group from C-4 carbon and hydroxyl group from C-3 or C-5 carbons from flavonoids structure (**Figure 2**). Further, aluminum can bond the orthodihydroxyl groups from A- and B-nucleus of flavonoids. The effect of the formation of these bonds results in coloration of the working solution in yellow due to resulted complex combinations.

Sample absorbances were measured in 10 mm cuvettes, at 510 nm, against redistilled water, and concentrations were calculated using the calibration curve drawn before each tests set, in the concentration range of 0.1–1 mg/mL of quercetin, used as reference flavonoid. Total flavonoids contents were expressed as mg quercetin equivalents per mL mushroom extract, and then converted to mushroom dry weight. Analytical data were collected on triplicate samples, mean values together with standard deviations were reported [69].

3.1.3 Antioxidant activity (AA)

Several chemical and biochemical assays can be used in order to evaluate the total antioxidant activity of mushrooms, and the 2,2-diphenyl-1-picrylhydrazyl DPPH$^•$ assay is one of the most frequently used [75–77]. Measurement principle is based on the fact that the antioxidant compounds from mushroom extracts release an electron or a hydrogen atom, and convert the DPPH$^•$ to a more stable, diamagnetic molecule, according to reaction below. DPPH$^•$ is a stable, long-lived organic nitrogen radical with a strong absorption around 517 nm (**Figure 3**).

The antioxidant activity of the studied mushroom extracts was assessed using DPPH$^•$ method. For good tests results, fresh ethanolic DPPH solutions (20 mg/mL) were prepared daily by weighing the necessary amount of DPPH powder (usually kept at −20°C), and kept in dark until experiments end. Samples were prepared by mixing aliquots of mushroom extract with DPPH solution, kept in dark at room temperature for 30 min, then sample absorbances were read to spectrophotometer, where zero absorbance was considered the extractant used for extracts preparation.

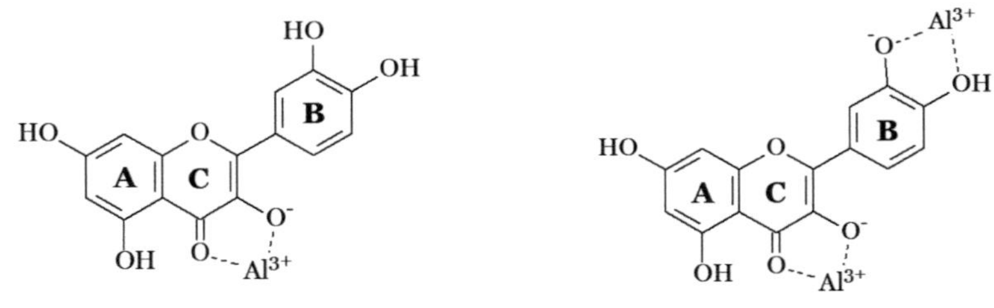

Figure 2.
Flavonoids complex combinations with Al^{3+}—quercetin example.

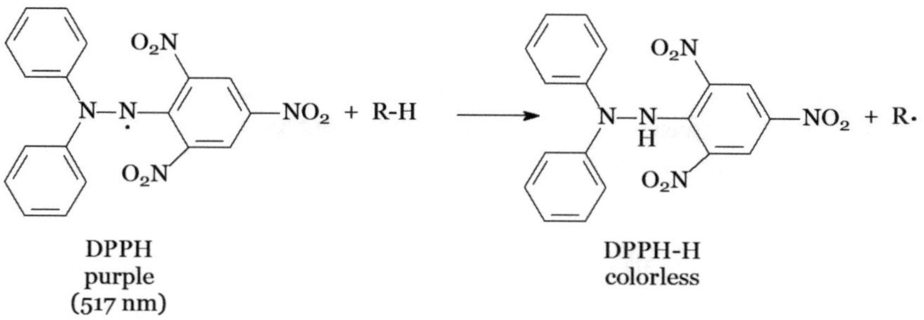

Figure 3.
Reaction scheme for antioxidant activity determination by DPPH method.

Reagent and sample blanks were prepared and measured for each test. Calculations were done according to equation:

$$AA\ (\%) = \left[A_{reagent\ blank} - \left(A_{extract} - A_{sample\ blank}\right) / A_{reagent\ blank}\right] \times 100 \qquad (1)$$

where AA is the global antioxidant activity of mushroom extract solutions, and A is the absorbance of the corresponding solution (as per subscripted text). As indicated by Eq. (1), results were calculated as % scavenging of DPPH at a fixed antioxidant concentration. A low absorbance of the tested sample indicates a high free-radical-scavenging activity.

3.1.4 Fourier transform infrared spectroscopy

To investigate the chemical functional groups of organic compounds in mushroom extracts, Fourier transform infrared spectroscopy was used. The chemical changes induced by extraction techniques as well as the various functional groups responsible for biological activities were detected in the mid-infrared absorption region using a Vertex 80v spectrometer (Bruker) equipped with a diamond attenuated total reflection crystal accessory [78]. The extracts were placed on the sample chamber of attenuated total reflection—Fourier transform infrared spectrometer without any preparation. The important absorption frequencies were noted in the range of 3600–600 cm^{-1}, as well as the fingerprint region of the spectra [79].

3.1.5 Inductive coupled plasma mass spectrometry (ICP-MS)

Minerals Cu, Fe, Zn, Mn are included in mushroom food chain, and, in low concentrations, they are considered antioxidant micronutrients. This designation is justified by their capability to catalyze some reactions producing reactive oxygen species, and their enzyme activation properties [10]. Edible mushrooms were reported as metals bio-accumulators, however high levels of essential metals intake could produce toxic effects when exceed certain values [78].

Trace elements Cu, Fe, Zn and Mn were measured by ICP-MS technique in aqueous solutions obtained by wet digestion. Before each test were performed the system calibration using Certipur® Certified Reference Material ICP multi-element standard IV (~1000 mg/L in 6.5% HNO_3, Merck). The instrumental parameters were: 1.5 kW plasma power, with 1 L/min argon nebulizer flow and 10.75 L/min plasma argon flow respectively, and precise analytical data were collected [10].

3.1.6 Peroxidases

Peroxidases are one of the classes of enzymes involved in the antioxidant defense mechanisms, together with superoxide dismutase, catalases, and others [10]. Experimental evaluation of peroxidase (POX) relies on its property to oxidize in the presence of hydrogen peroxide or other peroxide compounds (i.e. aromatics).

Oxidation of guaiacol by peroxidases in the presence of H_2O_2 is generally used for the colorimetric assay, absorbances measurements are performed at 420 nm, the chemical process involved is presented in **Figure 4**.

For accurate POX determination, fresh mushrooms are used to obtain the extracts that are further measured. Final results were reported as POX units per gram of mushroom. The unit of POX activity was defined as the oxidation of one micromole H_2O_2 per minute at 25°C (pH = 7.0).

3.1.7 Catalases

Catalases (CAT) are intracellular antioxidant enzymes present in edible mushrooms [10]. They are oxidoreductases, as they use hydrogen peroxide both as a receptor of electrons and as an electrons donor, decomposing it according to reaction presented in **Figure 5**.

Evaluation of catalase activity involves contacting a weighted amount of fresh mushroom with a measured volume of hydrogen peroxide at room temperature, allowed to stand for several minutes. The not-converted amount of hydrogen peroxide is then determined by titration with potassium permanganate in acid medium. Results are reported as CAT units per mushroom gram, while the unit of CAT activity is defined as the amount of enzyme decomposing one micromole H_2O_2 per minute at 25°C.

3.2 Analytical data and results interpretation

Total phenolic content of studied mushrooms (aqueous and hydroalcoholic extracts) is mentioned in **Figure 6**. As mentioned before, data were converted to milligrams of gallic acid equivalents (GAE) per gram of dried weight (d.w.). Experimental findings show certain differences between values for hydroalcoholic extracts and those prepared with water as solvent. On the other hand, in general for studied mushrooms, no significant differences between the two anatomic parts, as

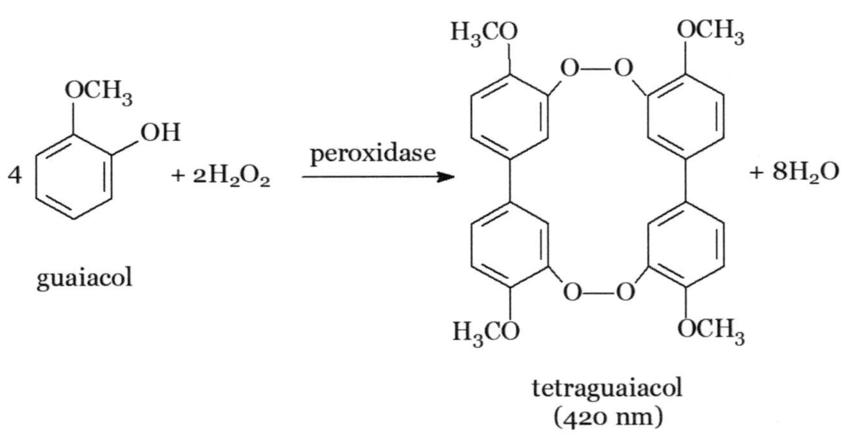

Figure 4.
Oxidation of guaiacol to tetraguaiacol, reaction catalyzed by peroxidase.

$$2H_2O_2 \xrightarrow{\text{catalase}} 2\,H_2O + O_2$$

Figure 5.
Decomposition of hydrogen peroxide catalyzed by catalase.

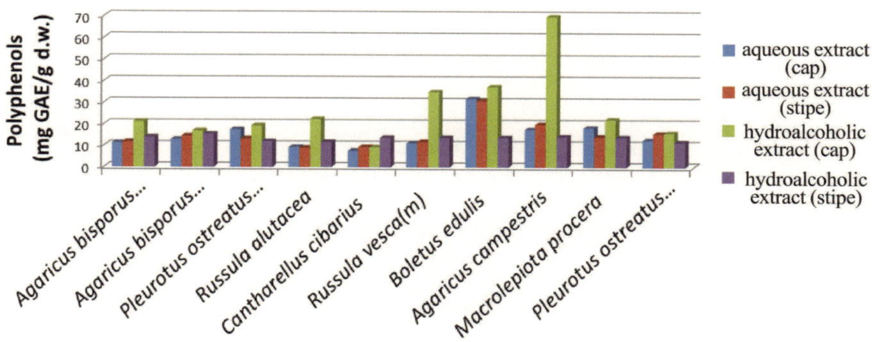

Figure 6.
Total content of polyphenols in studied mushrooms, aqueous and hydroalcoholic extracts, cap and stipe [mg GAE/g d.w.]

distinctly tested, cap and stipe. However, exceptions are observed, and will be further discussed. Concentration values found range between 9.28 ± 0.03 mg GAE/g d.w. (cap of *Cantharellus cibarius*) and 69.65 ± 0.23 mg GAE/g d.w. (cap of *Agaricus campestris*), average values being established for the cultivated species (**Figure 6**).

Considering both solvents and mushroom species, *Agaricus campestris* (cap) registered the highest difference between the content of phenolic compounds in the samples prepared in hydroalcoholic extractant instead of water, while the smallest one was determined in the case of *Macrolepiota procera* (stipe), who showed a slight preference for water. Intermediate differences were established for caps of *Russula vesca*, *Russula alutacea* and *Agaricus bisporus* white respectively, where the hydroalcoholic extractant was favorable to a better polyphenols extraction, while for *Boletus edulis* (stipe) water was a more convenient extractant for extraction of phenolic compounds.

Differences between anatomic parts were found to *Pleurotus ostreatus* (cultivated), *Russula alutacea*, *Boletus edulis* and *Macrolepiota procera*, higher total polyphenolic content was measured in caps than in stipes. An opposite behavior was found to *Cantharellus cibarius* mushroom hydroalcoholic extract, where TCP values were higher in stipe than in cap, and by 4.93 times of the case of *Agaricus campestris*. For aqueous extracts, closer values of TCP in caps and stipes were found. From the perspective of their origin, experimental findings for TCP showed lower average values for cultivated mushrooms than wild species group, regardless of the extractant type (15.03 mg GAE/g d.w. for water extracts and 20.14 mg GAE/g d.w. for hydro-alcoholic extracts). Also, no significant differences between caps and stipes for aqueous extracts, for both cultivated and wild species, average values for TCP were 13.54 mg GAE/g d.w. for cultivated and 13.05 mg GAE/g d.w. for wild ones respectively (excepting *Boletus edulis*).

Once the total flavonoid content (TFC) is considered, slight differences from the above mentioned findings were found. Thus, as may be observed in **Figure 7**, for some species (caps or stipes), flavonoids extraction in hydroalcoholic extractant was better than in water. TFC values measured in the hydroalcoholic extracts of *Russula alutacea* (cap), *Cantharellus cibarius* (stipe), *Russula vesca* (cap) and *Pleurotus ostreatus*—wild growing (cap), were higher than in their aqueous extracts. Aqueous extracts TFC exhibited values ranging between 0.22 ± 0.02 mg QE/g

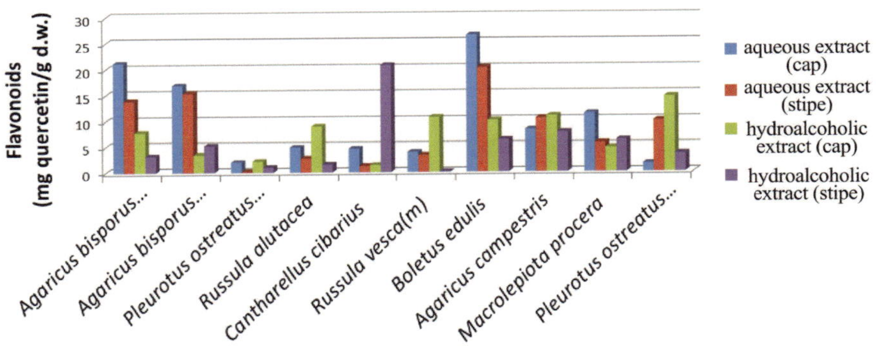

Figure 7.
Total flavonoids content in studied mushrooms, aqueous and hydroalcoholic extracts, cap and stipe [mg quercetin/g d.w.].

d.w. (stipe of *Pleurotus ostreatus* cultivated) and 26.51 ± 0.04 mg QE/g d.w. (cap of *Boletus edulis*), while TFC values for hydroalcoholic extracts were in the range of 0.12 ± 0.04 mg QE/g d.w. (stipe of *Russula vesca*) and 20.77 ± 0.06 mg QE/g d.w. (stipe of *Cantharellus cibarius*).

Similarities with total phenolics were found for total flavonoids detected, for comparisons made between mushroom species of different origin. Thus, experimental data showed that TFC average values in cultivated species were lower than in wild grown ones. Compared data for flavonoids found in caps and stipes showed that for both cultivated and wild species higher flavonoids content were noticed in cap for both extracts. Several exceptions have been noticed from this behavior: hydroalcoholic extracts of *Agaricus bisporus* brown (cultivated), and aqueous extract of *Agaricus campestris* and alcoholic extracts of *Cantharellus cibarius* and *Macrolepiota procera* respectively (wild species). One may conclude that total flavonoids content varied depending on the mushroom species and used extractant, polar solvents dissolving more flavonoids [69].

Analytical data for antioxidant activity of studied edible mushrooms extracts (cap and stipe), evaluated through DPPH method as previously described, showed some high and low limits. Thus, for aqueous extracts it was found that, *Agaricus bisporus brown* (cap) had the strongest DPPH radical-scavenging activity of 88.64%, while the lowest value of 25.72% was found in *Macrolepiota procera* (cap). When water-ethanol 50% (v/v) was used as extraction solvent, limit values were 74.93% for *Boletus edulis* (cap) and 13.61% respectively for *Russula alutacea* (stipe). Also, notable differences were found between analytical data recorded on cap and stipe of studied species [69]. Example of hydroalcoholic extracts is relevant: while most of mushroom species showed higher *AA*% values in caps than in stipes corresponding to same species, several exceptions were observed for *Cantharellus cibarius*, *Macrolepiota procera* and *Agaricus bisporus* brown where slight higher values were found in stipes than in caps. With regards to this phytochemical parameter (*AA*), a general behavior was noticed for studied mushrooms. Thus, notable differences between analytical data recorded for various species and when using the two extractant types.

By infrared spectroscopy several chemical functional groups that may be responsible for the antioxidant character of mushrooms, as quantified by classes of compounds or as a whole with the above mentioned ultraviolet-visible spectroscopic methods. Significant characteristic frequencies were observed in the range of 3600–600 cm^{-1} and fingerprint region, and were assigned to different organic compounds with —OH functional groups. As a relevant example, obtained results indicated that hydroalcoholic mushroom extracts may contain active functional groups as alcohols, esters and aldehydes [10].

Quantification of micronutrients in studied mushrooms showed, as may be observed in **Figure 8**, they are rather rich in Mn, Fe, Cu, and Zn, metals having a significant role in enzymatic systems activation. One may exemplify with data for species like *Boletus edulis* with Mn content of 130.73 mg/kg, and *Macrolepiota procera*, with Fe content of 715.14 mg/kg [10].

Enzymatic antioxidants peroxidase and catalase determinations in indigenous mushroom species were previously reported [10, 57], also **Figure 9** shows both enzymes activities. Significant variations were found for both studied enzymes. Higher values of catalase activity were found in species as *Agaricus bisporus white* and *brown* and *Russula vesca*, while species like *Boletus edulis* and *Pleurotus ostreatus wild* showed lower values. Measured values for catalase activity were in the range of 3.58–14.67 μmols H_2O_2/g/min. Also, highest values of peroxidase activity were found in mushroom species like *Russula alutaceea* and *Macrolepiota procera*, while lowest values of this enzyme were found in *Chantarellus cibarius*.

From the origin perspective, it was found that *Pleurotus ostreatus* cultivated had a 2.49 times higher catalase activity than the same wild species, while peroxidase activities for both wild and cultivated *Pleurotus ostreatus* were similar. Some correlations between metallic nutrients content enzymatic activities of mushrooms have been reported [10], and next chapter, through a chemometric approach will highlight further correlations between enzymatic and non-enzymatic antioxidant species, as were determined for studied mushrooms.

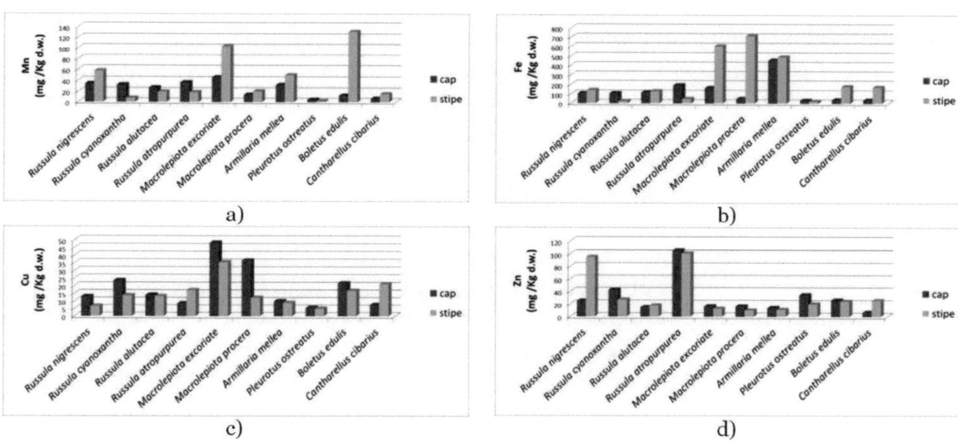

Figure 8.
Micronutrients concentrations in studied mushroom samples: a) Mn; b) Fe; c) Cu; d) Zn.

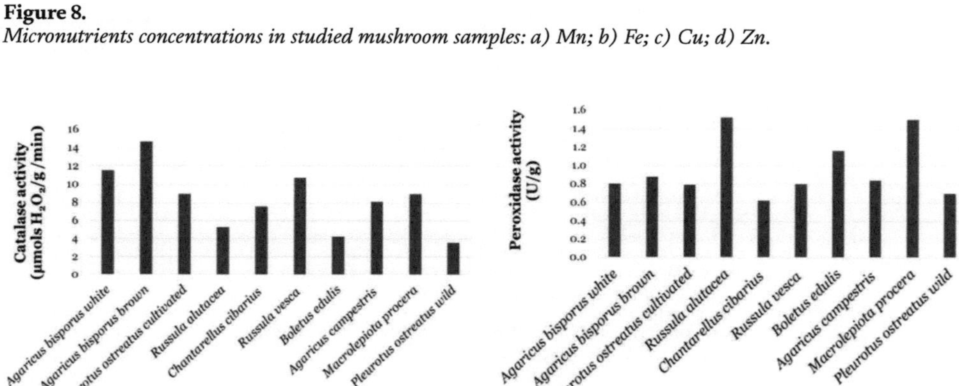

Figure 9.
Peroxidase and catalase activity in studied mushroom species.

4. Statistical analysis of data

Statistics have, through the descriptive methods of data analysis, powerful multidimensional analysis tools that can be used to design important information for fundamental research, applied research, market research, economic analysis, etc. Information can be hierarchized in terms of intensity of influence and can be analyzed as a whole and not independently [78].

One-way Analysis of Variance (ANOVA) was applied to the data set related to the ten mushrooms species, in order to observe whether there are any significant differences (Sig. < 0.05) between the means of the independent groups of variables. The produced F-statistic was higher for phenolics, flavonoids respectively antioxidant activity determined in aqueous extracts. Contrary, for the same parameters whose values were associated to hydroalcoholic extracts, the Sig. value higher than 0.05 indicated that there are no differences between groups in function of mushroom species.

In order to test the hypotheses of association between enzymatic and non-enzymatic antioxidants, the Bivariate (Pearson) Correlation was applied. From the large amount of information, the data shown in **Table 3** pointed out only the significant correlations, both for mushrooms' cap and stipe.

A strong positive relationship was observed between phenolics and flavonoids determined in aqueous extracts, regardless of the anatomic part of the mushrooms species. The strength of association was large but downhill between antioxidant activity determined in aqueous extracts and mushrooms species, if the last variable was defined in the next order: *Agaricus bisporus white*, *Agaricus bisporus brown*, *Pleurotus ostreatus* cultivated, *Russula alutacea*, *Chantarellus cibarius*, *Russula vesca*, *Boletus edulis*, *Agaricus campestris*, *Macrolepiota procera* and *Pleurotus ostreatus* wild respectively. Only for the stipe of the analyzed species, a strong relationship between the content of phenolics (in hydroalcoholic extracts) and the catalasic activity was determined. The relationship between catalase at least and mushroom species should be deeply analyzed, taking into account as much as possible types of potential linked variables, because in the last years the role of catalase (CAT), together with those of superoxide dismutase (SOD), was largely discussed in the context of the bioremediation biotechnologies.

No correlation could be observed for peroxidase, phenolics and antioxidant activity, irrespective of the extractant used and the anatomic part of the mushrooms. The Boxplot method was applied for graphically depicting groups of data related to phenolics, flavonoids and antioxidant activity, both in aqueous and hydroalcoholic extracts of mushrooms species. The quartiles of data represented in relationship

Anatomic part	Variables	Pearson's correlation coefficient
Cap	Phenolics—flavonoids (both in aqueous extracts)	0.857**
	Phenolics—antioxidant activity (both in hydroalcoholic extracts)	0.725*
	Antioxidant activity in aqueous extracts—mushrooms species	−0.643*
Stipe	Catalase—phenolics in hydroalcoholic extracts	0.700*
	Phenolics—flavonoids (both in aqueous extracts)	0.934**
	Antioxidant activity in aqueous extracts—flavonoids in hydroalcoholic extracts	−0.658*
	Antioxidant activity in hydroalcoholic extracts—flavonoids in aqueous extracts	0.679*

**Correlation is significant at the 0.05 level.*
***Correlation is significant at the 0.01 level.*

Table 3.
Pearson correlation coefficients for the analyzed variables.

with the anatomic part of the 10 species of mushrooms highlighted those species located far from the group in terms of their content in antioxidant compounds, respectively antioxidant capacity. The degree of dispersion and skewness in each category of data is indicated by the spaces between the different parts of the boxes. Variability outside the upper and lower quartiles of different variables was indicated by lines extending vertically from the boxes of the box plots (**Figures 10–12**).

Thus, both for cap and stipe, *Boletus edulis* remarked through a higher content of phenolic compounds in aqueous extract (**Figure 10a**). The median values for this parameter are close, independent of the anatomic part, a relative higher degree of dispersion being observed in the case of the mushrooms' cap than in the stipe. The variability of the data outside the upper quartile is however obvious for stipe. Also for the mushrooms' stipe, it cannot be about the scattering of the data if the phenolics in hydroalcoholic extracts are taken into account (**Figure 10b**). Based on the comparative analysis of the graphical representations (**Figure 10a** and **b**) it can be observed that the median value is superior for the group consisting of the caps of mushrooms in terms of phenolics in the case of the hydroalcoholic extracts than in the aqueous ones.

If the descriptive statistics was applied to content in flavonoids of the aqueous extracts of the mushrooms species—cap and stipe (**Figure 11a** and **b**), the second quartile for cap was close by the corresponding value for stipe, as it was observed for phenolics determined from aqueous extracts. The data sets show different trends for the other descriptive indicators. *Boletus edulis* (cap) and *Chantarellus cibarius* (stipe) are the mushroom species which detached from groups in terms of their content in flavonoids (in aqueous, respectively hydroalcoholic extracts).

Boxes and whisker plots quartiles cannot be overlapped if the antioxidant activity (both in aqueous and hydroalcoholic extracts) is analyzed **Figure 12**(**a** and **b**). The values determined for cap and stipe are higher in the case of the aqueous extracts of the mushrooms species, regardless of the quartiles displayed. The plotted outliers are not associated to a certain mushroom species.

Factorial analysis was applied to the enzymatic and non-enzymatic antioxidants of the mushroom species data sets. First of all, the matrix of the data correlation was developed and analyzed. The second phase of the study was based on the high values observed between some analyzed variables. Values of the total explained variance and Eigen values of the correlations matrix were generated. Two components were retained, the variables being represented on two factorial axis resulted from the combination of the initial variables (**Figure 13**).

After factors rotation (in order to obtain a better "angle" of view), PC1 explained 46.72% from the total variance, while PC2 explained 84.41%. It is thus possible to represent in the main plan the cloud of points. Two principal components were

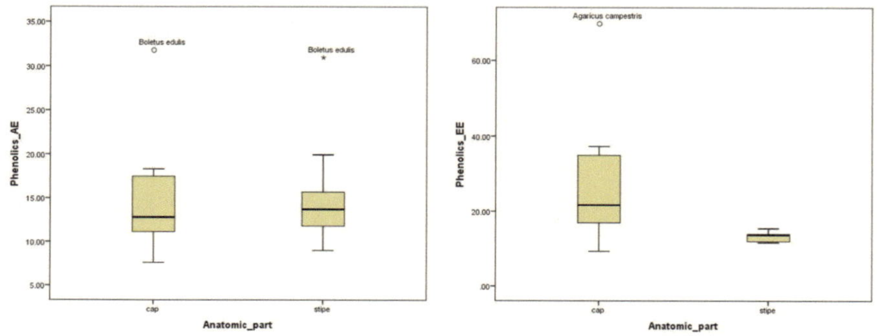

a b

Figure 10.
Box plots of phenolics in relationship with anatomic part of mushroom species: (a) aqueous extracts and (b) hydroalcoholic extracts.

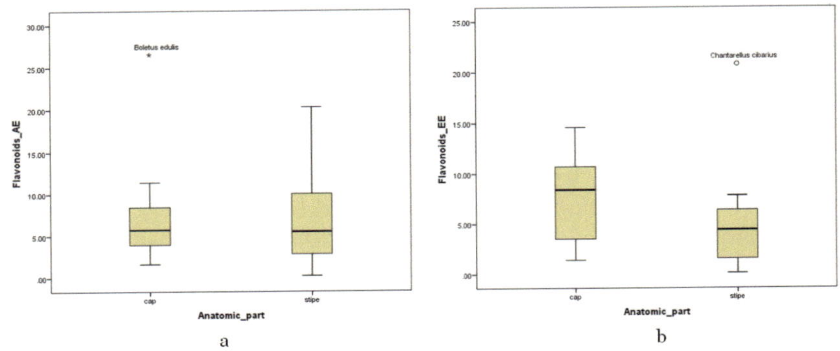

Figure 11.
Box plots of flavonoids in relationship with anatomic part of mushroom species: (a) aqueous extracts and (b) hydroalcoholic extracts.

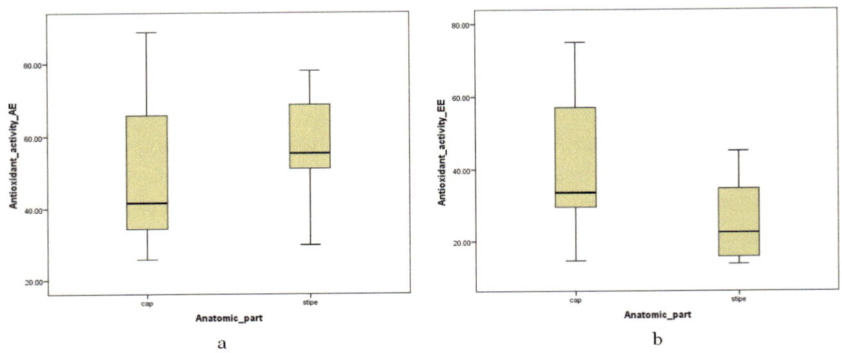

Figure 12.
Box plots of antioxidant activity in relationship with anatomic part of mushroom species: (a) aqueous extracts and (b) hydroalcoholic extracts.

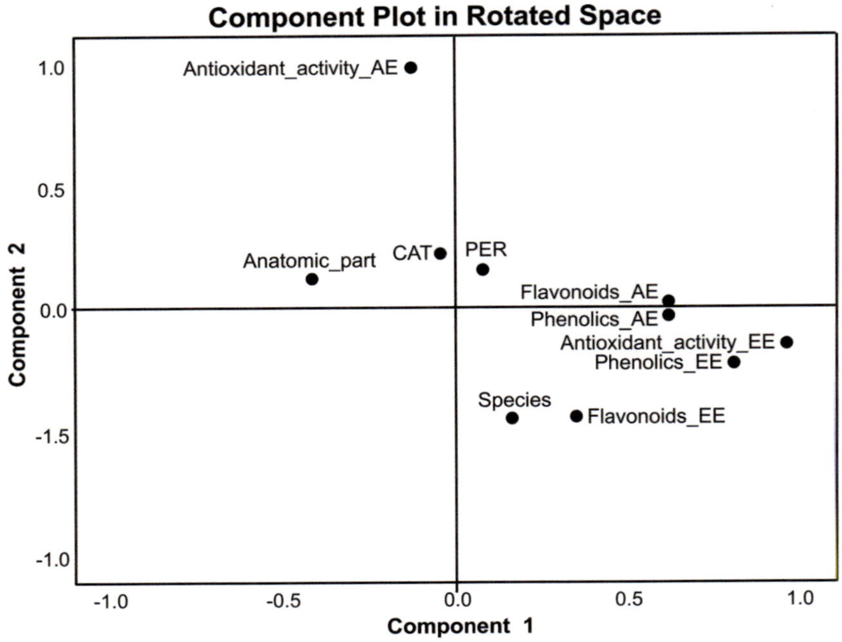

Figure 13.
Component plot in rotated space.

confirmed through PCA typical graphic representation, respectively the Screeplot (the Graphic of the eigenvalues). These two components (PC1 and PC2) obtained through axis rotation by Varimax method is represented in **Figure 13**. The values of the correlation coefficients (from matrix generated in the first step) are coordinates of the initial variables in the vectorial plan of the two principal components.

Concentration in phenolics and antioxidant activity (both in hydroalcoholic extract) were the major contributors to PC1, while the antioxidant activity of the mushroom species, determined in aqueous extracts, was the major contributor to PC2. The two factors can separate the area of antioxidants correlated with the anatomic part of the mushrooms by this one dominated by the same variables species dependent. The Screeplot and PC loadings suggests that the mushroom species affect mainly the antioxidant activity, determined in aqueous extract, while according to the contributors to PC1 is obvious that the anatomic part of the mushrooms influences the non-enzymatic antioxidants (phenolics, flavonoids in aqueous extracts) and antioxidant activity determined in hydroalcoholic extracts too. A linkage between the enzymatic antioxidants (catalase, peroxidase) and variables such as mushroom species and anatomic part was not observed by applying factorial analysis.

In order to group the datasets into similar data groups (classes, clusters), Hierarchical Cluster Analysis, who applies to small sets of data, was taken into account. The question arises as to whether in the set of variables there are identifiable groups, with similar characteristics, that characterize mushrooms' species (content of enzymatic and non-enzymatic antioxidant compounds). The square of the Euclidean distance was used to construct the matrix of similarities, while as method of aggregation—the Ward method. The clusters were formed considering the analyzed cases. All fungal species with similar characteristics (in terms of variables of interest) formed together clusters (**Figure 14**).

According to the antioxidants' concentration, in the initial stage of agglomeration different species of mushrooms and their anatomical parts form together three

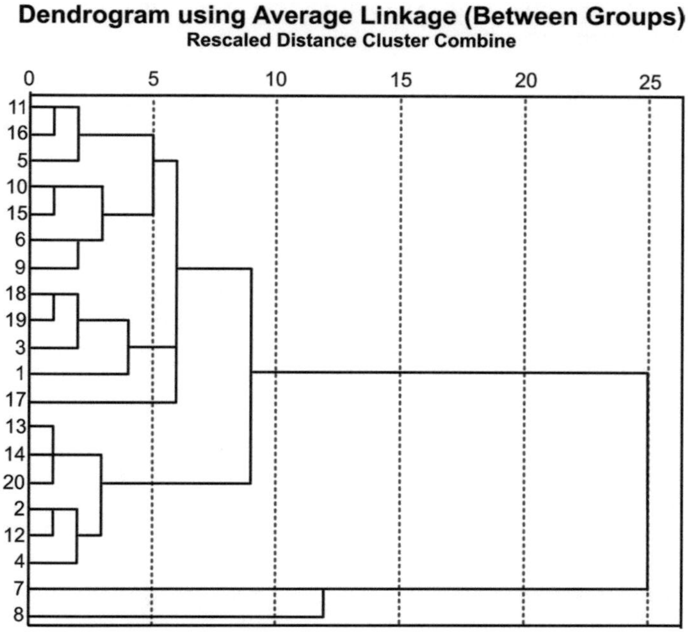

Figure 14.
Hierarchical cluster of the extracts: 1—A. bisporus white; 2—A. bisporus brown; 3—P. ostreatus cultivated; 4—R. alutacea; 5—C. cibarius; 6—R. vesca; 7—B. edulis; 8—A. campestris; 9—M. procera; 10—P. ostreatus wild (cap); 11—A. bisporus white; 12—A. bisporus brown; 13—P. ostreatus cultivated; 14—R. alutacea; 15—C. cibarius; 16—R. vesca; 17—B. edulis; 18—A. campestris; 19—M. procera; 20—P. ostreatus wild (stipe).

clusters. *B. edulis* (cap) and *A. campestris* (cap) remained isolated till the end stage of clusterization, being only ones clearly defined depending on the enzymatic and non-enzymatic antioxidants content. *A. bisporus brown* is the only species who aggregated in the initial stage as cap and stipe too. Excepting it, in the intermediate stages of the process the stipe of different mushroom species formed the first clusters, after that a mushroom cap joining to the structure already built. Finally, the clustering method leads to the formation of two clusters.

Chemometrics was applied in order to evaluate the traceability of Boletaceae mushrooms samples in combination with UV-visible and Fourier transform infrared (FTIR) spectroscopy [79], respectively in combination with inductively coupled plasma atomic emission spectrophotometer (ICP-AES), ultraviolet-visible (UV-Vis) and Fourier transform mid-infrared spectroscopy (FT-MIR) [80]. Through a chemometric approach were investigated the isotopic markers of *A. bisporus* origin [81] and the geo-traceability of mushrooms [82]. The Principal Component Analysis and Hierarchical Cluster Analysis were performed for fatty acids of *Ganoderma* species [83].

5. Conclusions

Within the analyzed group of autochtonous mushroom species, high concentration in phenolics and flavonoids were associated with the hydroalcoholic extracts. The mushrooms' anatomic part seemed to have influence on the concentration of non-enzymatic antioxidants, but only in the case of aqueous extracts. The antioxidant activity is species dependent, regardless of the type of mushroom extract.

Higher antioxidant abilities were determined for *Boletus edulis*, *Agaricus campestris* and *Chantarellus cibarius*. A significant correlation with the activity of catalase (CAT) was also established in the case of phenolic compounds. For these reasons at least these three mushroom species are promising in terms of designing functional foods and/or bioremediation processes. Chemometrics applied to heterogeneous data sets proved to be a powerful tool for selection of information and taking real time decisions in future research.

Author details

Cristiana Radulescu[1,2*], Lavinia Claudia Buruleanu[3], Andreea Antonia Georgescu[3] and Ioana Daniela Dulama[1]

1 Valahia University of Targoviste, Institute of Multidisciplinary Research for Science and Technology, Targoviste, Romania

2 Valahia University of Targoviste, Faculty of Sciences and Arts, Targoviste, Romania

3 Valahia University of Targoviste, Faculty of Environmental Engineering and Food Science, Targoviste, Romania

*Address all correspondence to: radulescucristiana@yahoo.com

References

[1] Chang ST. Mushroom research and development—Equality and mutual benefit. In: Royse DJ, editor. Mushroom Biology and Mushroom Products. University Park (PA): Pennsylvania State University; 1996

[2] Bielli E (trad. Gadei R). Cunoasterea, recunoasterea si cautarea celor mai cunoscute specii de ciuperci [Knowing, recognizing and searching for the most famous mushroom species]. Bucharest: All Educational; 1999. 320 p

[3] Busuioc G, Elekes CC, Stihi C. Ghid de caracterizare pentru unele specii de macromicete din flora spontana a judetului Dambovita [Characterization guide for some species of macromycetes in the spontaneous flora of Dambovita County]. Craiova: Sitech; 2011. 202 p

[4] Radulescu C, Stihi C, Popescu IV, Busuioc G, Gheboianu A, Cimpoca GV, et al. Determination of heavy metals content in wild mushrooms and soil by EDXRF and FAAS techniques. Ovidius University Annals of Chemistry. 2010;**1**(21):9-14

[5] Baldauf SL, Roger AJ, Wenk-Siefert I, Doolittle WF. A kingdom-level phylogeny of eukaryotes based on combined protein data. Science. 2000;**290**:972-977

[6] Chang ST, Buswell JA. Medicinal mushrooms—A prominent source of nutraceuticals for the 21st century. Current Topics in Nutraceutical Research. 2003;**1**:257-280

[7] Yasin H, Zahoor M, Yousaf Z, Aftab A, Saleh N, Riaz N, et al. Ethnopharmacological exploration of medicinal Mushroom from Pakistan. Phytomedicine. 2019;**54**(15):43-55

[8] Roupas P, Keogh J, Noakes M, Margetts C, Taylor P. The role of edible mushrooms in health: Evaluation of the evidence. Journal of Functional Foods. 2012;**4**:687-709

[9] Shin A, Kim J, Lim SY, Kim G, Sung MK, Lee ES, et al. Dietary mushroom intake and the risk of breast cancer based on hormone receptor status. Nutrition and Cancer. 2010;**62**:476-483

[10] Georgescu AA, Danet FD, Radulescu C, Stihi C, Dulama ID, Chelarescu DE. Determination of several elements in edible mushrooms using ICP-MS. Romanian Journal of Physics. 2016;**61**(5-6):1087-1097

[11] Srikram A, Supapvanich S. Proximate compositions and bioactive compounds of edible wild and cultivated mushrooms from Northeast Thailand. Agriculture and Natural Resources. 2016;**50**:432-436

[12] Adejumo TO, Awosanya OB. Proximate and mineral composition of four edible mushroom species from South Western Nigeria. African Journal of Biotechnology. 2005;**4**(10):1084-1088

[13] Akata I, Ergonul B, Kalyoncu F. Chemical Compositions and antioxidant activities of 16 wild edible mushroom species grown in Anatolia. International Journal of Pharmacology. 2012;**8**(2):134-138

[14] Kostić M, Smiljković M, Petrović J, Glamočlija J, Barros L, Ferreira ICFR, et al. Chemical, nutritive composition and a wide range of bioactive properties of honey mushroom *Armillaria mellea* (Vahl: Fr.) Kummer. Food & Function. 2017;**8**(9):3239-3249

[15] Yilmaz N, Turkekul I, Bulut S, Sahin F, Bayrak OF. Fatty acid composition in ten mushroom species collected from Middle Black Sea Region of Turkey. Asian Journal of Chemistry. 2013;**25**(3):1216-1220

[16] Kalač P. Chemical composition and nutritional value of European species of wild growing mushrooms: A review. Food Chemistry. 2009;**113**(1):9-16

[17] Pedneault K, Angers P, Gosselin A, Tweddell RJ. Fatty acid composition of lipids from mushrooms belonging to the family *Boletaceae*. Mycological Research. 2006;**110**(Pt 10):1179-1183

[18] Ribeiro B, Andrade PB, Silva BM, Baptista P, Seabra RM, Valento P. Comparative study on free amino acid composition of wild edible mushroom species. Journal of Agricultural and Food Chemistry. 2008;**56**(22):10973-10979

[19] Chen S, Oh SR, Phung S, Hur G, Ye JJ, Kwok SL, et al. Anti-aromatase activity of phytochemicals in white button mushrooms (*Agaricus bisporus*). Cancer Research. 2006;**66**:12026-12034

[20] Manzi P, Aguzzi A, Pizzoferrato L. Nutritional value of mushrooms widely consumed in Italy. Food Chemistry. 2001;**73**(3):321-325

[21] Spaulding T, Beelman RB. Survey evaluation of selenium and other minerals in Agaricus mushrooms commercially grown in the United States. Mushroom News. 2003;**51**:6-9

[22] Bailey RL, Dodd KW, Goldman JA, Gahche JJ, Dwyer JT, Moshfegh AJ, et al. Estimation of total usual calcium and vitamin D intakes in the United States. The Journal of Nutrition. 2010;**140**:817-822

[23] Poddar KH, Ames M, Hsin-Jen C, Feeney MJ, Wang Y, Cheskin LJ. Positive effect of mushrooms substituted for meat on body weight, body composition, and health parameters: A 1-year randomized clinical trial. Appetite. 2013;**71**:379-387

[24] Aras A, Khalid S, Jabeen S, Farooqi AA, Xu B. Regulation of cancer cell signaling pathways by mushrooms and their bioactive molecules: Overview of the journey from benchtop to clinical trials. Food and Chemical Toxicology. 2018;**119**:206-214

[25] Guarner F. Enteric flora in health and disease. Digestion. 2006;**73**(S1):5-12

[26] Radulescu C, Stihi C, Busuioc G, Gheboianu AI, Popescu IV, Cimpoca GV. Evaluation of essential and heavy metal levels in fruiting bodies of wild mushrooms and their substrate by EDXRF spectrometry and FAA spectrometry. Romanian Biotechnological Letters. 2010;**15**(4):5444-5456

[27] Persson H. Mushrooms. Medicine. 2012;**40**:135-138

[28] Stihi C, Radulescu C, Busuioc G, Popescu IV, Gheboianu A, Ene A. Studies on accumulation of heavy metals from substrate to edible wild mushrooms. Romanian Journal of Physics. 2011;**56**(1-2):257-264

[29] Busuioc G, Elekes C. Response of four *Russula* species under copper sulphate and lead acetate treatments. Notulae Botanicae Agrobotanici. 2013;**41**(2):538-545

[30] Radulescu C, Stihi C, Busuioc G, Gheboianu AI, Popescu IV. Studies concerning heavy metals bioaccumulation of wild edible mushrooms from industrial area by using spectrometric techniques. Bulletin of Environmental Contamination and Toxicology. 2010;**84**(5):641-647

[31] Popescu IV, Frontasyeva M, Stihi C, Cimpoca GV, Radulescu C, State G, et al. Atomic and nuclear methods applied in the study of heavy polluting elements. Romanian Reports in Physics. 2011;**63**:1205-1214

[32] Busuioc G, Elekes CC, Radulescu C, Gheboianu AI, Petrescu N, Serdin B.

The Influence of some heavy metals content of substrate and pH on bioabsorption level at six macromycetes species common in southern forest of Romania. Buletin USAMV—Agriculture Series. 2010;**67**(2):18-25

[33] European Commission—Directorate General for Research Communication Unit. Functional Foods. Luxembourg: Publications Office of the European Union; 2010. DOI: 10.2777/82512

[34] European Commission Concerted Action on Functional Food Science in Europe—FUFOSE. Scientific concepts of functional foods in Europe—Consensus document. British Journal of Nutrition. 1999;**81**(1):1-27

[35] Costin GM, Segal R. Alimente Functionale. Galati: Academica; 1999. 207 p

[36] Reis FS, Martins A, Vasconcelos MH, Morales P, Ferreira ICFR. Functional foods based on extracts or compounds derived from mushrooms. Trends in Food Science & Technology. 2017;**66**:48-62

[37] Ferreira ICFR, Barros L, Abreu RMV. Antioxidant in wild mushrooms. Current Medicinal Chemistry. 2009;**16**:1543-1560

[38] Soković M, Glamočlija J, Ćirić A, Petrović J, Stojković D. Mushrooms as sources of therapeutic foods. In: Holban AM, Grumezescu AM, editors. Therapeutic Foods. Vol. 8. Cambridge: Academic Press; 2018. pp. 141-178

[39] Kalaras MD, Richie JP, Calcagnotto A, Beelman RB. Mushrooms: A rich source of the antioxidants ergothioneine and Glutathione. Food Chemistry. 2017;**233**:429-433

[40] Alves MJ, Ferreira ICFR, Froufe HJC, Abreu RMV, Martins A, Pintado M. Antimicrobial activity of phenolic compounds identified in wild mushrooms, SAR analysis and docking studies. Journal of Applied Microbiology. 2013;**115**:346-357

[41] Helleno SA, Barros L, Martins A, Morales P, Fernandez-Ruiz V, Glamoclija J, et al. Nutritional value, bioactive compounds, antimicrobial activity and bioaccessibility studies with wild edible mushrooms. Food Science and Technology. 2015;**63**:799-806

[42] Cheung PK. Mini-review on edible mushrooms as source of dietary fiber: Preparation and health benefits. Food Science and Human Wellness. 2013;**2**:162-166

[43] Ma G, Yang W, Zhao L, Pei F, Fang D, Hu Q. A critical review on the health promoting effects of mushrooms nutraceuticals. Food Science and Human Wellness. 2018;**7**:125-133

[44] Putnik P, Bursác Kovačević D, Režek Jambrak A, Barba F, Cravotto G, Binello A, et al. Innovative "green" and novel strategies for the extraction of bioactive added value compounds from citrus wastes—A review. Molecules. 2017;**22**:680

[45] Putnik P, Lorenzo JM, Barba FJ, Roohinejad S, Jambrak AR, Granato D, et al. Novel food processing and extraction technologies of high-added value compounds from plant materials. Food. 2018;**7**:106

[46] Roselló-Soto E, Parniakov O, Deng Q, Patras A, Koubaa M, Grimi N, et al. Application of non-conventional extraction methods: Toward a sustainable and green production of valuable compounds from mushrooms. Food Engineering Reviews. 2016;**8**(2):214-234

[47] Corrêa RCG, Barros L, Fernandes A, Sokovic M, Bracht A, Peralta RM, et al. A natural food ingredient based on ergosterol: Optimization of the extraction from *Agaricus blazei*,

evaluation of bioactive properties and incorporation in yogurts. Food & Function. 2018;**9**(3):1465-1474

[48] Moumita S, Das B, Hasan U, Jayabalan R. Effect of long-term storage on viability and acceptability of lyophilized and spray-dried synbiotic microcapsules in dry functional food formulations. LWT—Food Science and Technology. 2018;**96**:127-132

[49] Nowak R, Nowacka-Jechalke N, Juda M, Malm A. The preliminary study of prebiotic potential of Polish wild mushroom polysaccharides: The stimulation effect on Lactobacillus strains growth. European Journal of Nutrition. 2018;**57**:1511-1521

[50] Francisco CRL, Heleno SA, Fernandes IPM, Barreira JCM, Calhelha RC, Barros L, et al. Functionalization of yogurts with *Agaricus bisporus* extracts encapsulated in spray-dried maltodextrin crosslinked with citric acid. Food Chemistry. 2018;**245**:845-853

[51] Kurt A, Gençcelep H. Enrichment of meat emulsion with mushroom (*Agaricus bisporus*) powder: Impact on rheological and structural characteristics. Journal of Food Engineering. 2018;**237**:128-136

[52] Lu X, Brennan MA, Serventi L, Liu X, Guan W, Brennan CS. Addition of mushroom powder to pasta enhances the antioxidant content and modulates the predictive glycaemic response of pasta. Food Chemistry. 2018;**264**:199-209

[53] Khan AA, Gani A, Masoodi FA, Mushtaq U, Naik AS. Structural, rheological, antioxidant, and functional properties of β-glucan extracted from edible mushrooms *Agaricus bisporus*, *Pleurotus ostreatus* and *Coprinus attrimentarius*. Bioactive Carbohydrates and Dietary Fibre. 2017;**11**:67-74

[54] Lavelli V, Proserpio C, Gallotti F, Laureati M, Pagliarini E. Circular reuse of bio-resources: The role of Pleurotus spp. in the development of functional foods. Food & Function. 2018;**9**:1353-1372

[55] Gil-Ramírez A, Morales D, Soler-Rivas C. Molecular actions of hypocholesterolaemic compounds from edible mushrooms. Food & Function. 2018;**9**:53-69

[56] Lee SR, Roh HS, Lee S, Park HB, Jang TS, Ko YJ, et al. Bioactivity-guided isolation and chemical characterization of antiproliferative constituents from morel mushroom (*Morchella esculenta*) in human lung adenocarcinoma cells. Journal of Functional Foods. 2018;**40**:249-260

[57] Dissanayake AA, Zhang CR, Mills GL, Nair MG. Cultivated maitake mushroom demonstrated functional food quality as determined by in vitro bioassays. Journal of Functional Foods. 2018;**44**:79-85

[58] Akata I, Zengin G, Picot CMN, Mahomoodally MF. Enzyme inhibitory and antioxidant properties of six mushroom species from the Agaricaceae family. South African Journal of Botany. 2018. DOI: 10.1016/j.sajb.2018.01.008. https://www.sciencedirect.com/science/article/pii/S0254629917314850

[59] Perera PK, Li Y. Mushrooms as a functional food mediator in preventing and ameliorating diabetes. Functional Foods in Health and Disease. 2011;**1**(4):161-171

[60] Aisala H, Laaksonen O, Manninen H, Raittola A, Hopia A, Sandell M. Sensory properties of nordic edible mushrooms. Food Research International. 2018;**109**:526-536

[61] Ishara JRM, Sila DN, Kenji GM, Buzera AK. Nutritional and functional properties of mushroom (*Agaricus bisporus & Pleurotus ostreatus*) and their blends with maize flour. American

Journal of Food Science and Technology. 2018;**6**(1):33-41

[62] Politowicz J, Lech K, Sánchez-Rodríguez L, Figiel A, Szumny A, Grubor M, et al. Volatile composition and sensory profile of oyster mushroom as affected by drying method. Drying Technology. 2018;**36**:685-696

[63] Muhammad AI, Liao X, Cullen PJ, Liu D, Xiang Q, Wang J, et al. Effects of nonthermal plasma technology on functional food components. Comprehensive Reviews in Food Science and Food Safety. 2018;**17**:1379-1394

[64] Kozarski M, Klaus A, Jakovljevic D, Todorovic N, Vunduk J, Petrovic P, et al. Antioxidants of edible mushrooms. Molecules. 2015;**20**:19489-19525

[65] Lushchak VI. Free radicals, reactive oxygen species, oxidative stress and its classification. Chemico-Biological Interactions. 2014;**224**:164-175

[66] Georgescu AA, Danet AF, Radulescu C, Stihi C, Dulama ID, Buruleanu CL. Nutritional and food safety aspects related to the consumption of edible mushrooms from Dambovita County in correlation with their levels of some essential and non-essential metals. Revista de Chimie. 2017;**68**(10):2402-2406

[67] Folin O, Ciocalteu V. Tyrosine and tryptophan determinations in proteins. The Journal of Biological Chemistry. 1927;**73**:627-650

[68] Agbor GA, Vinson JA, Donnelly PE. Folin-Ciocalteau reagent for polyphenolic assay. International Journal of Food Science, Nutrition and Dietetics. 2014;**3**(8):147-156

[69] Buruleanu LC, Radulescu C, Georgescu AA, Danet FA, Olteanu RL, Nicolescu CM, et al. Statistical characterization of the phytochemical characteristics of edible mushroom extracts. Analytical Letters. 2018;**51**(7):1039-1059

[70] Amic D, Davidovic-Amic D, Beslo D, Rastija V, Lucic B, Trinajstic N. SAR and QSAR of the antioxidant activity of flavonoids. Current Medicinal Chemistry. 2007;**14**:827-845

[71] Farkas O, Jakus J, Heberger K. Quantitative structure-antioxidant activity relationship of flavonoid compounds. Molecules. 2004;**9**:1079-1088

[72] Marin FR, Frutos MJ, Perez-Alvarez JA, Martinez-Sanchez F, Del Rio JA. Flavonoids as nutraceuticals: Structural related antioxidant properties and their role on ascorbic acid preservation. Studies in Natural Products Chemistry. 2002;**26**(G):741-778

[73] Kozarski M, Klaus A, Vunduk J, Zizak Z, Niksic M, Jakovljevic D, et al. Nutraceutical properties of the methanolic extract of edible mushroom Cantharellus cibarius (Fries): Primary mechanisms. Food & Function. 2015;**6**(6):1875-1886

[74] Kamtekar S, Keer V, Patil V. Estimation of phenolic content, flavonoid content, antioxidant and alpha amylase inhibitory activity of marketed polyherbal formulation. Journal of Applied Pharmaceutical Science. 2014;**4**(9):61-65

[75] Akar Z, Kucuk M, Doğan H. A new colorimetric DPPH• scavenging activity method with no need for a spectrophotometer applied on synthetic and naturalantioxidants and medicinal herbs. Journal of Enzyme Inhibition and Medicinal Chemistry. 2017;**32**(1):640-647

[76] Olteanu RL, Nicolescu CM, Bumbac M. Influence of phytochemical reductive capacity on ultraviolet–visible

spectroscopic behavior of silver nanoparticles. Analytical Letters. 2017;**50**(17):2786-2801

[77] Radulescu C, Stihi C, Ilie M, Lazurca D, Gruia R, Olaru OT, et al. Characterization of phenolics in lavandula angustifolia. Analytical Letters. 2017;**50**(17):2839-2850

[78] Buruleanu L, Radulescu C, Georgescu AA, Nicolescu MC, Olteanu RL, Dulama ID, et al. Chemometric assessment of the interactions between the metal contents, antioxidant activity, total phenolics, and flavonoids in mushrooms. Analytical Letters; in press. DOI: 10.1080/00032719.2018.1528268

[79] Yao S, Li T, Liu HG, Li JQ, Wang YZ. Traceability of Boletaceae mushrooms using data fusion of UV–visible and FTIR combined with chemometrics methods. Journal of the Science of Food and Agriculture. 2018;**98**(6):2215-2222

[80] Qi L, Liu H, Li J, Li T, Wang Y. Feature fusion of ICP-AES, UV-Vis and FT-MIR for origin traceability of boletus edulis mushrooms in combination with chemometrics. Sensors. 2018;**18**(1):241

[81] Chung IM, Han JG, Kong WS, Kim JK, An MJ, Lee JH, et al. Regional discrimination of *Agaricus bisporus* mushroom using the natural stable isotope ratios. Food Chemistry. 2018;**264**:92-100

[82] Tokul-Olmez O, Kaplaner E, Ozturk M, Ullah Z, Duru ME. Fatty acid profile of four Ganoderma species collected from various host trees with chemometric approach. Biochemical Systematics and Ecology. 2018;**78**:91-97

[83] El Sheikha AF, Hu DM. How to trace the geographic origin of mushrooms? Trends in Food Science & Technology. 2018;**78**:292-303

Chapter 3

Alternatives for Sugar Replacement in Food Technology: Formulating and Processing Key Aspects

Marko Petković

Abstract

The physical, chemical, thermal, rheological, and sensory characteristics of spreads with noncarbohydrate nutritive sweeteners (such as polyols), produced on ball mill, could be predicted. Spreads with 70 and 100% maltitol, as a sweetener, produced on different temperatures (30, 35, 40°C) and mixer speed rotations (60, 80, 100 r/min), give the spreads with very good or excellent sensory characteristics, characteristic spreadability without sandiness (gritty texture), good melting behavior, and pleasant taste. Both process parameters are very important and have the dual effect on spread quality. The best spread quality, considering all characteristics, has the spread with 100% maltitol, produced on the highest process parameters (40°C, 100 r/min).

Keywords: polyols, sugar, confectionery, spread, chocolate

1. Introduction

Low-energy foods, or products with reduced energy value, are very popular among the consumers. There is a need for developing the new and enriched existing products nowadays. The consumers' awareness is raised. They want to primarily satisfy the need for the sweet taste, as well as for maintaining or reducing your body weight, without any consequences for your health [1].

Replacement of nutritive sweeteners with other low-energy sweeteners (such as polyols) can change the sensory characteristics of the basic product. Proper selection of raw materials, as well as proper management of the technological process, can obtain the products with optimal sensory properties [2].

This chapter should explain the possibility of native sugar substitution with sugar alternatives, such as polyols (maltitol, mannitol, sorbitol, xylitol, isomalt, lactitol, erythritol).

Polyols are the most suitable nutritive sugar substitute for confectionery products, such as chocolate, chocolate desserts/bars, spreads and cocoa cream products, hard/soft candies or chewing gums, bakery products, and nonalcoholic beverages. Choosing an adequate polyol, as a sugar (sucrose) nutritive substitute, gives the possibility for a product that has almost unchanged sensory properties and that has maintained a sweet taste. This information is of crucial importance for industrial production. It provides important technological parameters and information for changing the process parameters and the need for new equipment:

1. Polyols are very stable at high temperatures and do not react with amino acids. Generally, polyols [3] have about 40–50% less energy value than sucrose that has significantly more stability than monosaccharides from which they are produced because they do not have a carbonyl group.

2. Do not participate in caramelization and Maillard reactions.

3. Do not ferment in the oral cavity and therefore do not cause caries having a pleasant and neutral taste.

4. Give the feeling of cooling, especially sorbitol and xylitol.

Polyols are normally present in little amounts in organic products and in addition to specific sorts of vegetables or mushrooms. They are additionally recognized as safe food additives [4, 5].

In addition, polyols are used as emulsifiers, stabilizing agents, flavor enhancer humectant, moisture binding, controlling crystallization, anticaking agent, bulking agent, cryoprotectors, etc. According to the European Union regulation, polyols are nutritive food additives and identified by E number, i.e., sorbitol (E420), mannitol (E421), isomalt (E953), maltitol (E965), lactitol (E966), xylitol (E967), and erythritol (E968). Polyols must be always listed in the ingredient lists on the food package, and its use in food products is defined by the Regulation (EC) 1333/2008 on food additives [6].

The acceptable daily intake (ADI) dose of polyols has not been defined. Polyols are marked to be *quantum satis* level for all purposes [6]. But polyols have a few side effects when overeaten, such as laxative effect, gastrointestinal symptoms, bloating, diarrhea, and abdominal pain. Therefore, if any food product containing more than 10% added polyols must include the statement "excessive consumption may produce laxative effects" [7, 8]. So, polyols are helpful in weight control, diabetes, and tooth decay [9, 10].

2. The basic physical and chemical properties of polyols

Polyols (sugar alcohols) are nutritive sweeteners obtained by the catalytic hydrogenation of the oxo-group of natural sugars, i.e., by substituting an aldehyde or keto group with hydroxyl [11].

The sweetness of sugar alcohols (polyols) is shown in **Table 1** [3–7].

The sweetness of polyols is lower than sucrose. Therefore, polyols might be used as a bulk sweetener. The desired level of sweetness and flavor of food products are achieved by the combination of polyols and non-nutritive, usually artificial, sweeteners. Polyols are responsible for texture, preservation, filling, moisture capture, and cooling effect in the mouth [5, 11–14]. Polyol sweetness, such as maltitol, is up to 90% of the sucrose sweetness [8].

In addition, consumption of products containing polyols does not increase the glucose level in blood or insulin secretion, and thus food products with polyol are recommended for people with diabetes. Polyols are alike prebiotics and can normalize, as fibers, intestine function [4, 7, 8]. Polyols, such as maltitol, are able to increase mineral bioavailability in humans and rats [15].

DOI: http://dx.doi.org/10.5772/intechopen.82251

Polyol	Energy value (kJ/g)			Glycemic index	Sweetness	Hygroscopicity	Heat of solution (kJ/kg)	Cooling effect (kcal/g)	Melting point (°C)	Solubility (g/100 g H_2O (25° C))	Heat stability (°C)	Acid stability	Molecular weight (g/mol)	Molecular formula
	EU*	USA**	Japan											
Xylitol	10.0	10.0	16.7	13	1.0	High	−153	Very cool (−36.6)	94	63	>160	2–10	152.2	$C_5H_{12}O_5$
Maltitol	10.0	8.8	8.4	35	0.9	Median	−79	/(−18.9)	150	60–65	>160	2–10	344.3	$C_{12}H_{26}O_{12}$
Sorbitol	10.0	10.8	12.5	9	0.6	Median	−111	Cool (−26.5)	97	70–75	>160	2–10	180.2	$C_6H_{14}O_6$
Erythritol	0***	0***	0***	0	0.6	Very low	−180	Cool (− 18.9)	126	37–43	>160	2–10	122.1	$C_4H_{10}O_4$
Mannitol	10.0	6.7	8.4	0	0.6	Low	−121	Cool (−28.9)	165	18–22	>160	2–10	182.2	$C_6H_{14}O_6$
Isomalt	10.0	8.4	8.4	9	0.5	Low	−39	/(−9.4)	145–150	25–28	>160	2–10	344.3	$C_{12}H_{26}O_{12}$
Lactitol	10.0	8.4	8.4	6	0.4	Median	−53	Slightly cool (−13.9)	122	55–57	>160	2–10	344.3	$C_{12}H_{26}O_{12}$ $C_{12}H_{26}O_{12} \cdot H_2O$
Sucrose	16.7	16.7	16.7	68	1.0	Median	−18	/(−4.3)	190	67	160–186	>3	342.3	$C_{12}H_{22}O_{11}$

*-European Union.
**-United States of America.
***-0–0.8368 kJ/g.

Table 1.
The basic important physical and chemical parameters of different polyols.

The energy value of polyols and glycemic index, in relation to sucrose, are shown in **Table 1** [3–7].

Polyols have a lower nutritional value (10 kJ/g) than sugars (16.7 kJ/g, **Table 1**), due to slower and incomplete absorption from the intestine. The results of polyol fermentative degradation by the intestinal flora are fatty acids and gases [7, 10]. Due to their incomplete absorption, polyols produce a lower glycemic response than carbohydrates (**Table 1**) and therefore might be useful in diabetic diets, causing smaller increases in blood glucose and insulin levels as compared to sugar and other carbohydrates [8].

Molecular weight and melting point (**Table 1**) are good to investigate when initially screening ingredients for applications. Maltitol is, for example, suitable as a bulking agent, without an additional agent to be needed. The solubility of a polyol can lead to recrystallization in a product. It is important to adapt the polyol to the specific application and monitor throughout shelf life of the final product (e.g., to predict the shelf life).

The primary application of polyols in foods is shown in **Table 2** [2, 16–18].

When the first formulas for sweet products were developed, several facts of sweetener choice had to be taken into account (**Table 2**). Obviously, the choice between crystalline polyol and liquid polyol (polyol syrup) will depend on the type of product and the ability to mix them or on the type of carbohydrate sweetener to be replaced. When we compare the physical and chemical properties of sucrose with polyols, and we talk about chocolate or spread, the most optimal choice of sucrose substitute is maltitol [2, 16–18].

Polyols, such as maltitol, affect seeding technique (β_V stable cocoa butter crystal) and rheological, textural, and thermal characteristics of dark chocolates [19]. Temper index value (TIV) gives information about tempering degree of the chocolates. TIV values of dark chocolate with sucrose were as TIV values of dark chocolate with maltitol. Dark chocolate with maltitol should satisfy the required terms of demoulding process, sensory characteristics (color, appearance, texture), thermal behavior (melting demands), and shelf life stability [20]. Particle size distribution and texture of dark chocolate with maltitol did not change strongly. Thermal characteristics, such as melting, were determined by DSC method (differential scanning calorimeter). The sweetener concentration, as well as seeding, didn't change melting characteristics at all [19]. Rheological properties are described by the flow curves where the shear stress of the sugar-free chocolates is a function of shear rate. The shear rate versus viscosity indicates the shear behavior of the sugar-free dark chocolate. This variation between the flow behaviors can be generally affected by

Polyol	Food application
Xylitol	Jellies, chewing gums, coatings for gum, mint-flavore candies
Maltitol	Chocolate, spread, hard candies, chewing gums, coating for gums
Sorbitol	Chewing gums, tablets, candies, humectants, plasticizers, hard candies, baked goods
Erythritol	Hard/soft candies, chocolate, beverages, bakery products, chewing gums
Mannitol	Dusting power, chewing gums, effervescent products
Isomalt	Chewing gum, dusting powder
Lactitol	Candies, frozen desserts, jams and jellies, chocolate, dusting powder, bulking agent, baked products

Table 2.
The primary application of different polyols.

ingredients (type and concentration of fat, sweetener, emulsifier) and process parameters (refining process, such as refining time, temperature, mixer speed rotation, etc.) [19–22].

The use of maltitol, as the only sweetener in the production of chocolate and spread products, does not require the use of other non-nutritive artificial sweeteners. Non-nutritive artificial sweeteners are calorie less. Only aspartame provides 4 kcal/g but is consumed in small amounts (about 200 times sweeter than sucrose) and contributes negligible energy [14].

Application of polyols in confectionery products, such as cookies, shows that maltitol has a similar effect due to their comparable molecular weights. Cookie characteristics with maltitol are similar to the cookies with sucrose, with a crumb structure, comparable rise and greater diameter increase, higher hardness, and brittleness [23, 24]. The crust lightness for maltitol cookies was decreased by 25% because Maillard reactions were not occurred [25]. Cookies with maltitol have a significantly softer texture too. When we analyze the relative sweetness of cookies, maltitol cookies were comparable to cookies with sucrose, and general acceptance of cookies with maltitol was significantly higher [24].

Semidried jerky made by polyols enhanced the quality attributes, especially xylitol, which is very appropriate in meat composition. The increase in the level of polyols causes a slight reduction in the pH values, regardless of the polyol type. The water activity of semidried meat jerky with polyols led to lower water activity and depends on the molecular size of polyols. As the molecular weight of polyols is larger, its solution has a greater osmotic pressure than the same amount of sucrose solution [26]. Sugar alcohols, such as sorbitol and xylitol, make the metal-chelating ability and cell reinforcement movement, reducing the oxidation of meat products. Kim et al. demonstrated that sorbitol increased the textural characteristics of pork meat jerky [27].

Flavor release in chewing gum depends of the type and particle size of polyol. Particle size distribution of polyols was determined by modern laser diffraction technique using a Malvern Mastersizer. As the particle size of the polyols is decreased, the surface area for flavor release is increased. The distribution of highly polar flavor compounds, such as the high-intensity sweeteners (HIS), is higher too because the high-intensity sweeteners are less entrapped by the gum base during manufacture. But, some flavor compounds had a higher flavor release when formulated with a larger particle size of polyols, specifically limonene [28]. Optimal dimensions of polyol particles in the production of chewing gum are sorbitol 200 μm, mannitol 60 μm, xylitol 90 μm, and maltitol 35 μm [11].

Replacement of nutritive sweeteners with other low-energy sweeteners can change the textural and sensory characteristics of the basic confectionery product, such as spreads or chocolate. Proper selection of raw materials, as well as proper management of the technological process, can obtain the final products of optimal sensory properties.

2.1 The physical and chemical properties of maltitol

Maltitol (E 965, 4-O-α-d-glucopyranosyl-d-glucitol) is a white crystalline powder, odorless, and not enzyme-resistant. It is produced from starch, by hydrogenating maltose or a very high maltose glucose syrup and crystallization from the maltose syrup [29, 30]. Maltitol is a disaccharide (equal parts of glucose and sorbitol) which causes a mild cooling effect, with physicochemical characteristics similar to sucrose. As a sucrose substitute, the technological parameters of chocolate and spread production do not need to be changed. As well as the other polyols, maltitol is able to change the rheological characteristics of spread and chocolate [6, 8]. Its use is like a bulking agent, sweetener, emulsifier, humectants, stabilizer, and thickener. As a fat substitute, maltitol gives a creamy texture to food [30–32].

Maltitol is very soluble in water (66 g/100 g at 25°C) and has a higher solubility than sucrose at temperatures >40°C. Due to its low hygroscopicity and high-temperature stability, it is used in many baked products as a bulk sweetener and energy reducer [10]. The ADI value of maltitol is not specified and depends of consumer organism condition, its age, sex, etc. The minimum amount exceeding 25–30 g/kg body weight per day can reveal laxative effect; thus, the maximum amount should not exceed 50 g total [13, 33, 34]. Maltitol is enzyme-sensitive and slowly digested in the small intestine (absorption range is from 5 to 80%) to glucose and sorbitol, where the nonabsorbed part passes to the colon where it undergoes fermentation by bacteria. It does not undergo Maillard reactions (browning process) and caramelization, with negligible cooling effect with other polyols [30–32, 35]. Maltitol has a low glycemic index, increases the mineral bioactivity in humans, and reduces postprandial glycemic responses with short-chain fructo-oligosaccharides [15, 30–32, 36, 37].

3. The basic spread-/chocolate-making process in laboratory

The basic spread/chocolate-making process in laboratory is outlined in **Figure 1**.

Spreads, unlike chocolate, do not contain cocoa butter (which requires a conching and tempering phase, **Figure 1**) but special vegetable fats, so spread production is cheaper and less demanding. Special vegetable fats have emphasized plastic properties, such as palm fat. At the conventional method, chocolate needs a special tempering procedure to satisfy texture, quality, and appearance [38–40].

In laboratory conditions, laboratory ball mill is most often used for spread refining (particle size reduction, the largest particles should be below 30 μm, optimum size is 20–22 μm). The process parameters are the temperature, the mixer speed rotation, the diameter/number of balls, the speed of spread recirculation, the

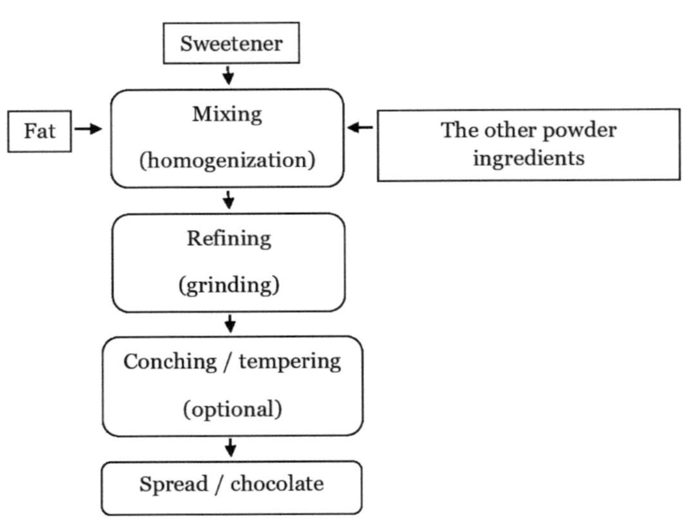

Figure 1.
The basic spread-/chocolate-making process in laboratory.

fat/solid content, the water content, the type/amount of emulsifiers, and the particle size distribution [18, 41].

Spreads were produced by a nonconventional producing method, in laboratory ball mill (capacity 5 kg). Laboratory ball mill is a horizontal or vertical cylinder, with a double wall and a bottom. Hot water circulates through the wall and bottom. In the central part of the cylinder, there is a shaft with a mixer and blades. The interior of the laboratory ball mill (60–80%) is filled with stainless steel balls (9.1 mm diameter, 30 kg weight). The speed of spread recirculation is 10 kg/h. The refining (grinding) time is 150 minutes. Experimental spread samples (50 g) were sealed in plastic glasses and stored at a temperature of 25°C [17].

The basic ingredients for spreads are sweetener 47.4%, palm fat 36%, cocoa powder 7%, whole milk powder 7%, soy flour 2%, lecithin 0.5%, and flavor 0.1%. The used sweeteners are maltitol (100%), sucrose (100%), and a combination of maltitol and sucrose (70/30% and 30/70% ratio). Spreads were produced at different temperatures (30, 35, 40°C) and mixer speed rotations (clockwise—60, 80, 100 r/min). Spread with 100% maltitol, produced at temperature 30°C and 60 r/min, is labeled as "M–30–60" [17]. The fat content is over 32%; there are a few changes in yield value with any further additions [20, 42]. The temperature ratio was chosen because the solid palm fat content on temperatures over 30°C is less than 1%. Higher-temperature range causes higher energy costs too.

The refining time (150 minutes) was purposely chosen. It provides the absence of sandiness (particles >30 μm).

4. The basic physical and chemical characteristics of spreads

4.1 The chemical composition of spreads

The chemical composition of spreads was determined by methods [12]: total carbohydrates (polarimetry), total fat (Soxhlet), total proteins (Kjeldahl), total moisture (thermogravimetry), total sucrose (polarimetry), and total maltitol (HPLC).

Spreads with 100% maltitol have the lowest energy value (20.37 kJ/g—100% maltitol; 21.42—70% maltitol; 24.29—100% sucrose; 23.28—70% sucrose) [17]. This result is expected [17, 18, 20]. Spreads with sucrose have slightly increased moisture content (1.06%, the moisture content of spreads with maltitol is 0.73–0.78) because sucrose has pronounced hygroscopic properties compared to maltitol, which is in acceptable limit [17, 42].

4.2 Particle size distribution of spreads

Particle size distribution was determined by the microscopic method [17].

Spreads with 100% maltitol have lower parameters of medium dimension of largest particles (61.67–70.58 μm—100% maltitol; 62.76–64.5 μm—70% maltitol; 73.12–88.55 μm—100% sucrose; 69.16–72.43 μm—100% sucrose). Sucrose is more hygroscopic and partly recrystallizes and forms the agglomerates. The mixer speed rotation is more dominant; the higher speed rotations affect the stronger frictional forces and smaller dimensions of the largest particles. Spreads, produced on maximum speed rotation (100 r/min), have the lowest average values of the largest particles [17]. Chocolate with a high percentage of particles above 30 μm has a gritty or coarse perception in the mouth [17, 43, 44].

4.3 Textural characteristics of spreads

Textural characteristics of spreads were determined on the Texture Analyzer TE32 by the manufacturer's specified method TA Chocolate spread_SPRD2_SR [10, 12]. The experimental results define the following parameters: the firmness (the maximum force at the curve of the force dependence of time) and the work of shear (determined by the area under the curve, which defines the spreadability of spreads). The firmness and the work of shear are outlined in **Figure 2**.

Spreads with maltitol (70, 100%) have harder crystals (the maximum penetration force 1093–1351 g, **Figure 2**) because of higher crystalline strength [2, 10]. The hardness of solid tempered chocolate is correlated with the type of fat and its content, the particle size distribution, the type of sweetener, and the tempering process [17, 43–45]. But, replacement of maltitol as a bulking agent in the study of Konar had no substantial effect on chocolate hardness [46].

When the process parameters are increased, the firmness is slightly increased, while the spreadability is decreased, regardless of the sweetener type. So, these parameters are in high correlation (R^2 = 0.927, 0.953, 0.989, 0.961). The mixer speed rotation is a dominant. Combination of sucrose and maltitol results in parameter variation.

The application of higher values of process parameters makes it possible to obtain more fine solid particles, a homogenous mass with a wider specific surface area. This area contributes better suspension of continuous fat phase.

4.4 Rheological characteristics of spreads

Rheological characteristics of spreads were determined on the HAAKE RheoStress 600 rotary viscometer (temperature 40 ± 0.1°C, the shear rate 0–60/s, the shear stress 0.1–10 Pa, frequency 1 Hz (ω = 6.28 rad/s)) [17]. Dynamic oscillatory measurements are applied to monitor the modulus of elasticity G' and the modulus of viscosity G," which are determined in the linear viscoelastic regime (LVE).

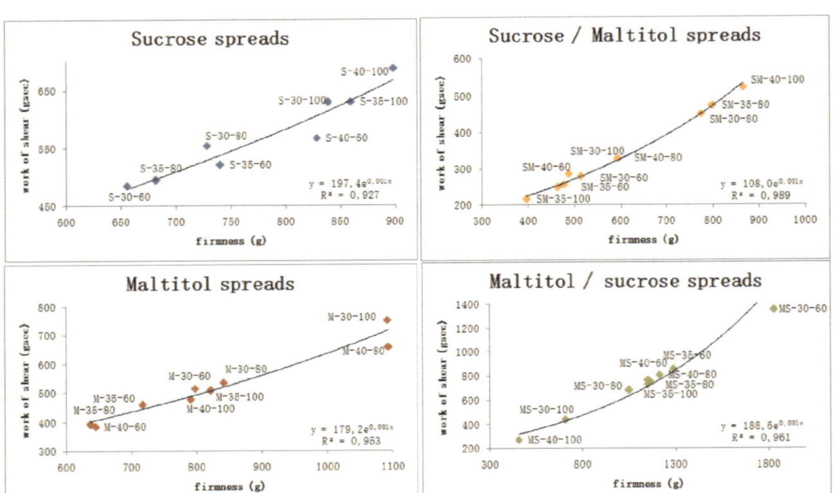

Figure 2.
The firmness and the work of shear of spreads.

Spreads are classified into pseudoplastic systems in which the solid particles (sucrose, maltitol, cocoa particles, whole milk powder, etc.) are wrapped into a fat continuous phase and fitted in it. The thixotropic loop, which occurs on behalf of viscous and elastic areas, is described by the yield stress t (Pa), Casson plastic viscosity η (Pas), and thixotropic area P (Pa/s). The parameters of viscosity are outlined in **Table 3**. Flow curves were produced by standard procedure and graphically outlined in **Figure 3**. Casson plastic viscosity range of spreads with maltitol is variable and between 1.218 and 558.3 Pa/s, as a result of maltitol and sucrose combination. Different hygroscopicity of polyols causes different Casson viscosity of chocolate; chocolate with high levels of polyols has higher Casson plastic viscosity [42].

The plastic viscosity of spreads with maltitol is larger than with sucrose (**Table 3**, [2, 16–18]). Casson plastic viscosity for chocolate is between 2.1 and 3.9 Pas [20, 42]. The viscosity of spreads with maltitol in our study is in this range and Casson yield values too. The sweetener type is more dominant than the mixer speed rotation; maltitol makes the yield stress decrease, and viscosity and thixotropic area are increased. The higher plastic viscosity of spreads with maltitol might be in a correlation with its slightly lower density (1.60 g/cm^3) than sucrose (1.63 g/cm^3) [34]. Temperature is more dominant than the mixer speed rotation; increasing the temperature, viscosity, yield stress, and thixotropic area become

Spread	Yield stress (τ) (Pa)	Thixotropic area (P) (Pa/s)	Casson plastic viscosity η (Pas)	Spread	Yield stress (τ) (Pa)	Thixotropic area (P) (Pa/s)	Casson plastic viscosity η (Pas)
S-30-60	7.38	575.8	0.912	SM-30-60	2.76	479.40	2.73
S-30-80	8.56	447.7	1.806	SM-30-80	4.08	542.10	2.68
S-30-100	8.53	374.0	1.216	SM-30-100	4.07	263.40	1.218
S-35-60	7.57	401.9	1.216	SM-35-60	3.46	499.00	3.099
S-35-80	6.34	220.0	1.597	SM-35-80	6.42	497.00	3.018
S-35-100	8.72	401.3	172.800	SM-35-100	2.87	23302.00	1.425
S-40-60	8.25	364.6	1.794	SM-40-60	43.32	451.70	2.693
S-40-80	7.29	241.0	1.430	SM-40-80	6.13	481.70	3.001
S-40-100	9.32	361.8	150.600	SM-40-100	5.56	669.30	4.493
M-30-60	5.56	2099.00	7.137	MS-30-60	11.68	2229.00	14.094
M-30-80	7.85	1181.00	4.717	MS-30-80	7.17	837.90	4.907
M-30-100	9.49	2436.00	4.061	MS-30-100	6.18	643.10	5.166
M-35-60	3.29	1547.00	558.300	MS-35-60	9.15	1570.00	8.614
M-35-80	8.13	1469.00	6.030	MS-35-80	5.99	682.20	4.73
M-35-100	6.40	1715.00	5.748	MS-35-100	9.03	1049.00	7.004
M-40-60	2.96	1177.00	449.000	MS-40-60	8.18	818.80	4.795
M-40-80	11.68	2342.00	13.636	MS-40-80	34.30	999.50	6.015
M-40-100	5.53	1613.00	5.672	MS-40-100	8.09	549.20	1.821

Table 3.
The parameters of viscosity of spreads.

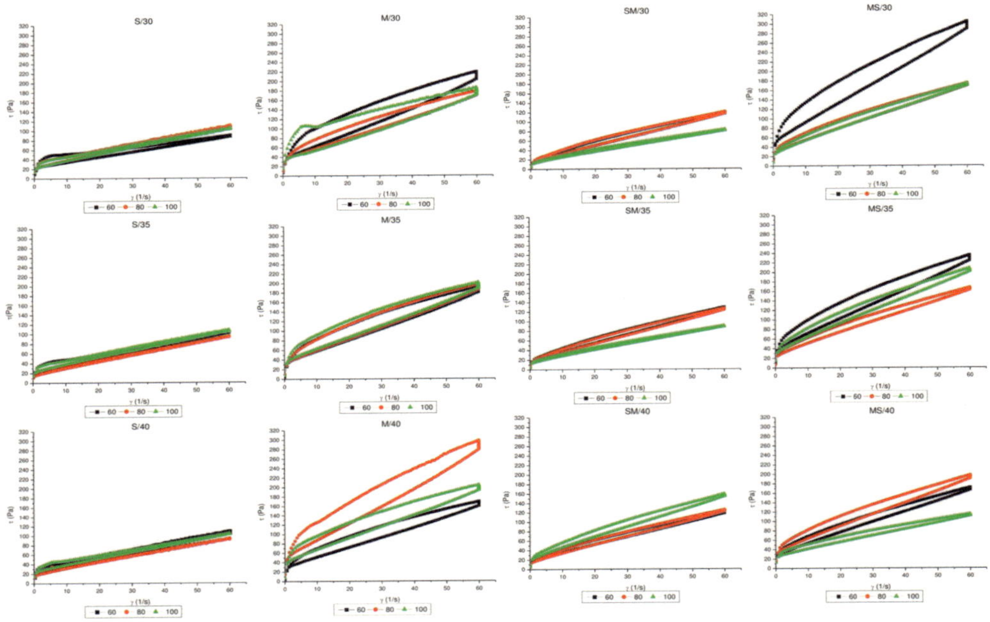

Figure 3.
Flow curves of spreads.

lower. High values of plastic viscosity can be explained with a theory that particle size distribution becomes wider with a heterogeneous specific surface area. Smaller particles fill spaces between larger and reduce the viscosity [17, 20, 42]. Casson yield values for spreads with maltitol are within the limits of the parameters for milk chocolate and have been reported to be between 2 and 18 Pas [13]. In general, chocolate with a high level of maltitol (75%) has a very similar flow index as chocolate with sucrose [45].

Spreads with 100% maltitol have a higher thixotropic area (**Figure 3**). This behavior can be mainly connected with the high molecular mass of maltitol [42]. Maltitol crystals, after refining, are coarse and have heterogeneous distribution of particle size with a large specific surface area. The high Casson yield value of maltitol products is a consequence of agglomeration in mass. The high molecular mass of maltitol increases the nonpolar intermolecular interactions. As a result, the mass becomes firmer and agglomerated, and thus more energy is required to start the flowing process [45]. The sucrose/maltitol combination spreads cause a huge variation in rheology results.

The parameters of loss coefficient (tanδ, $\tan\delta = G''/G'$) are shown in **Table 4** [2, 16–18].

The area of the elastic component is for $\tan\delta > 1$, while the viscous area is for $\tan\delta < 1$ (**Table 4**). Spreads belong to viscoelastic systems, in which the elastic component of the system is dominant. This characteristic is very important in process design and quality assessment for food such as butter or spreads [47]. For spreads with a point of intersection, the viscous area is dominant on frequencies below 2 Hz, up to the point of intersection, where the elastic area becomes more dominant in the system [2, 16–18].

The dominant process parameter is the mixer speed rotation. The Casson plastic viscosity decreases with an increase of shear rate (mixer speed rotation). Chocolate with 100% maltitol was found to be very similar to the control (chocolate with sucrose) in the tested plastic viscosity [45].

Spread	tanδ	Spread	tanδ
S-30-30	0.833880	SM-30-30	0.499218
S-30-40	0.855628	SM-30-40	0.533216
S-30-50	0.884045	SM-30-50	0.551405
S-35-30	1.002739	SM-35-30	0.865578
S-35-40	0.837005	SM-35-40	0.795029
S-35-50	1.022986	SM-35-50	0.697446
S-40-30	0.870694	SM-40-30	0.916253
S-40-40	0.961491	SM-40-40	0.781274
S-40-50	1.027589	SM-40-50	0.678345
M-30-30	0.492289	MS-30-30	0.540614
M-30-40	1.019242	MS-30-40	0.786911
M-30-50	1.359460	MS-30-50	0.750791
M-35-30	1.162145	MS-35-30	0.460028
M-35-40	1.079023	MS-35-40	0.793502
M-35-50	1.072265	MS-35-50	0.778090
M-40-30	0.942098	MS-40-30	0.754004
M-40-40	1.164598	MS-40-40	0.814258
M-40-50	0.949932	MS-40-50	0.826978

Table 4.
The loss coefficient parameters of spreads.

4.5 Thermal characteristics of spreads

Thermal characteristics of ingredients and spreads are analyzed by TG analysis device "LECO TG701." The thermal decomposition of spreads is monitored in the air stream (3.5 dm^3 min^{-1}), with the heating rate of 5°C min^{-1} in the temperature range of 25–800°C [2, 10, 16–18]. The peak of DTG curves gives the temperature corresponding to maximum degradation (T_{max}). The characteristic peaks of sweeteners and spreads are outlined in **Table 5**. Thermal decomposition of sucrose, maltitol, and palm fat is outlined in **Figure 4**.

Thermal decomposition of maltitol and sucrose is two-phase (**Figure 4**). The initial thermal decomposition of maltitol starts at 269.79°C with the distinguished peak at 340.39°C (**Table 6**). The mass loss in the first thermal decomposition phase occurs rapidly without complex biopolymer and is about 80%. The second decomposition phase occurs slower. The second peak of maltitol decomposition is 481.317° C (the residual mass at 600°C is 0.9%). The peaks of maltitol are sharper and with more expressed the inflection point.

Thermal decomposition of spreads is outlined in **Figures 5** and **6**. Spreads with 100% maltitol have the initial peak of maltitol decomposition, which is lower and between 335 and 356°C (**Table 6**), as a result of the presence and similar initial temperature decomposition of palm fat. The peak of palm fat decomposition is between 387 and 406°C. This peak similarity disables the mass loss determination. The mixer speed rotation is dominant; higher mixer speed rotation parameter makes the peak formation on higher temperatures.

	Sucrose peak, T_{max}/°C	Maltitol peak, T_{max}/°C	Palm fat peak, T_{max}/°C		Sucrose peak, T_{max}/°C	Maltitol peak, T_{max}/°C	Palm fat peak, T_{max}/°C
Sucrose ($T1_{max}$/°C)	240.793			Sucrose ($T2_{max}$/°C)	493.023		
Maltitol ($T1_{max}$/°C)		340.391		Maltitol ($T2_{max}$/°C)		481.317	
Palm fat ($T1_{max}$/°C)			402.784	Palm fat ($T2_{max}$/°C)			/
S-30-60	231.74		354.27	MS-30-60	219.61	326.40	395.56
S-30-80	233.97		378.51	MS-30-80	221.86	328.98	397.36
S-30-100	236.27		347.97	MS-30-100	223.96	331.66	399.55
S-35-60	216.42		371.95	MS-35-60	226.18	334.44	401.64
S-35-80	219.21		352.36	MS-35-80	228.44	337.02	381.07
S-35-100	221.75		354.53	MS-35-100	230.81	339.54	394.60
S-40-60	223.93		346.01	MS-40-60	232.89	342.08	386.85
S-40-80	226.01		348.28	MS-40-80	235.01	344.39	387.26
S-40-100	228.14		361.57	MS-40-100	221.45	346.52	391.35
M-30-60		338.38	387.29	SM-30-60	217.31	332.55	379.04
M-30-80		341.70	390.21	SM-30-80	219.73	334.83	381.29
M-30-100		344.32	392.81	SM-30-100	221.82	336.49	383.67
M-35-60		346.85	395.14	SM-35-60	224.13	339.18	385.83
M-35-80		349.22	397.12	SM-35-80	226.35	341.53	387.79
M-35-100		351.75	399.41	SM-35-100	228.74	/	389.89
M-40-60		354.38	401.89	SM-40-60	231.18	/	392.19
M-40-80		356.83	405.01	SM-40-80	233.39	325.73	394.34
M-40-100		334.83	406.08	SM-40-100	235.67	327.61	385.47

$T1_{max}$/°C, temperature peak of the first phase; $T2_{max}$/°C, temperature peak of the second phase.

Table 5.
The characteristic peaks of sweeteners and spread ingredients.

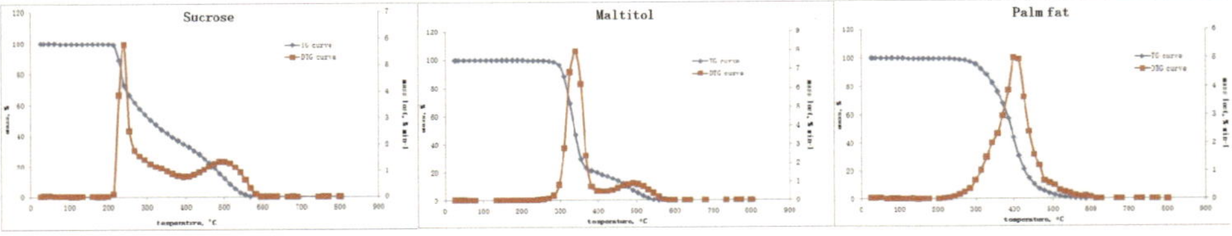

Figure 4.
TG curves and DTG curves of sucrose, maltitol, and palm fat.

Spreads with 70% sucrose/30% maltitol and 30% sucrose/70% maltitol have the variations of peak visibility (**Figure 6**). The result of these variations is actually the peak overlapping due to the final thermal decomposition of sweetener and initial thermal decomposition of palm fat. The other reasons of peak overlapping are inhomogeneous structure, the refining process, and the presence of emulsifier

Quality factor	Impact factor	Temperature (°C)								
		30			35			40		
Mixer speed rotation (o/min)		60	80	100	60	80	100	60	80	100
						S				
External appearance	0.6	2.4	1.8	1.5	1.5	1.5	1.5	1.5	1.5	2.1
Texture	0.8	3.2	2.8	2.4	2.8	2.8	2.4	2.8	2.8	3.6
Chewiness	1	4	4	4	4	4	4	4.5	4.5	4.5
Flavor	0.6	3	2.7	3	2.7	3	2.4	3	3	3
Taste	1	4	4	4.5	4.5	4.5	4.5	4.5	4.5	5
Σ		16.6	15.3	15.4	15.5	15.8	14.8	16.3	16.3	18.2
Quality category		VG	VG	VG	VG	VG	VG	VG	VG	E
Mixer speed rotation (o/min)						M				
External appearance	0.6	2.1	2.7	2.7	2.4	2.7	3.0	2.1	2.4	2.4
Texture	0.8	3.2	3.6	4.0	2.8	3.6	3.6	3.2	3.6	4
Chewiness	1	4	4	4	3.5	4	4.5	4	4.5	4.5
Flavor	0.6	2.4	2.7	2.4	2.4	3	3	2.7	2.7	2.7
Taste	1	4	4	4.5	4	5	5	5	5	5
Σ		15.7	17.0	17.6	15.1	18.3	19.1	17.0	18.2	18.6
Quality category		VG	VG	E	VG	E	E	VG	E	E

Table 6.
Sensory evaluation of spread quality using the scoring procedure.

(lecithin). This peak was observed on 230°C by analyzing spreads with maltitol (**Figures 5** and **6**), close to temperature decomposition of sucrose. Maltitol spreads do not contain sucrose; the peak decomposition of cocoa powder, soya powder, and whole milk is on temperature over 500°C, so this peak belongs to lecithin. The peak of lecithin thermal decomposition is 200°C [18, 48].

The refining process and ingredient decomposition make the peak become lower and wider. In general, the most stable spreads with 100% maltitol and sweetener combination are produced on higher process parameters. The peak position and temperature can be used to detect the unknown spread ingredient, comparing it to known peaks.

4.6 Sensory characteristics of spreads

Sensory analysis of spreads was made 7 days after the stabilization. The scoring method of the five-member panel evaluated the following parameters of quality (score 0–5, **Tables 6–8** [2, 16–18]): the external appearance (the shape, the color, and the structure), the texture, the chewiness, the taste, and the flavor (aroma). Score is multiplied by the appropriate impact factor to calculate the points. The sum of points defines the quality category: excellent (E), very good (VG), good (G), sufficient, (S), and insufficient (I) [17].

The total score of all spreads in this study ranged from 15.1 to 19.1 gives the spreads with very good and excellent quality (**Tables 7** and **8**). Spreads with maltitol (100 and 70%) have a better structure and external appearance. However, the addition of maltitol has a negative effect on flavor (spreads with maltitol have less pronounced flavor). Increasing the process parameters, excellent sensory

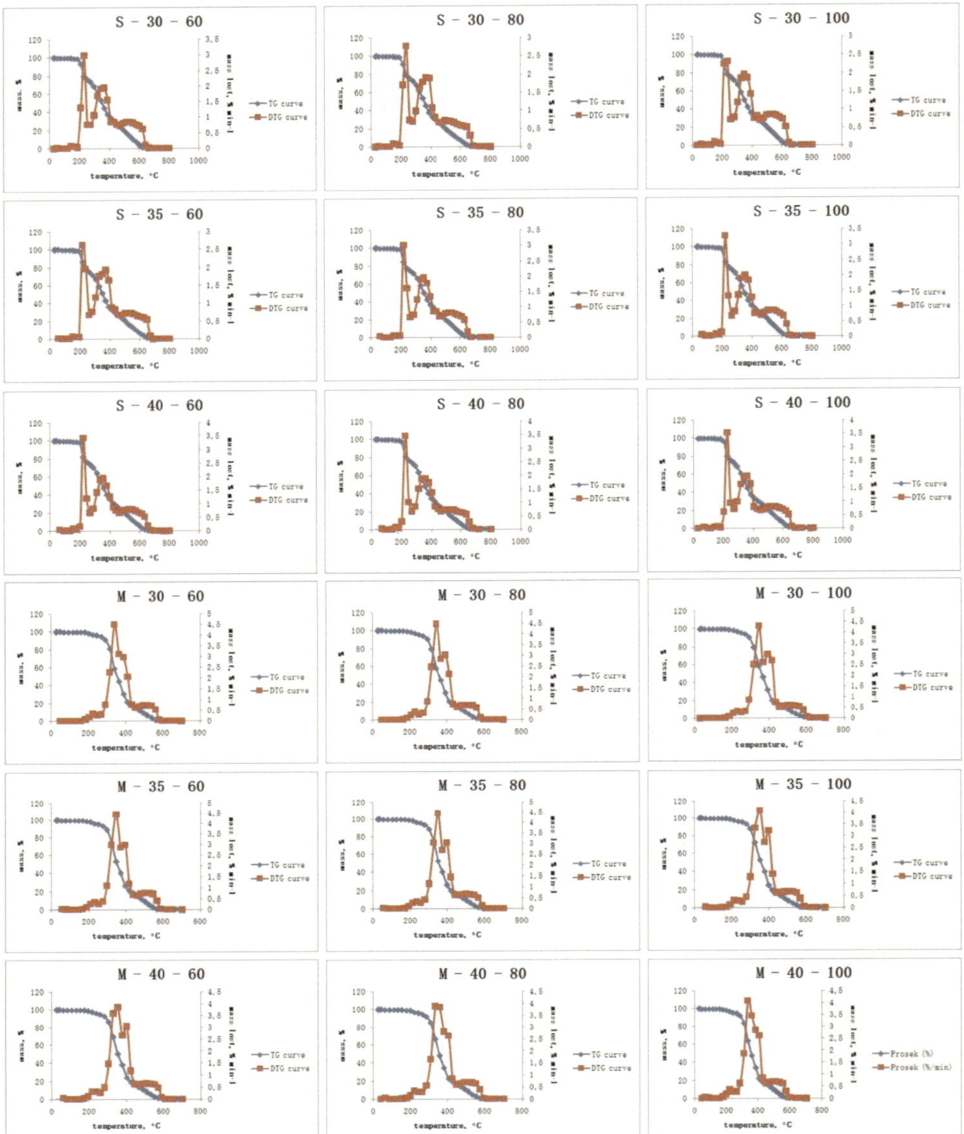

Figure 5.
TG curves and DTG curves of spreads with 100% sucrose and 100% maltitol.

properties of spreads with maltitol are achieved. The hardness of maltitol spread texture is higher which is not good (high hardness, less spreadability).

Qualitative data analysis (QDA) method consists of evaluating (from 1 to 5) individual quality elements and their input into polar coordinates. On each polar coordinate, there are five labels. By merging the labels marked with individual quality elements, a quality diagram of the entire spread is obtained (**Figure** 7). This method is highly suitable for monitoring the quality of products in regular production [17, 45]. As the diagram area is larger, the quality category of spread is larger too.

It was noticed that spreads with 100 and 70% maltitol have less pronounced flavor and slightly bitter and fruity taste (as a secondary sensory characteristic) [2]. Flavor, taste, and chewiness are the most dominant quality factors, with the average

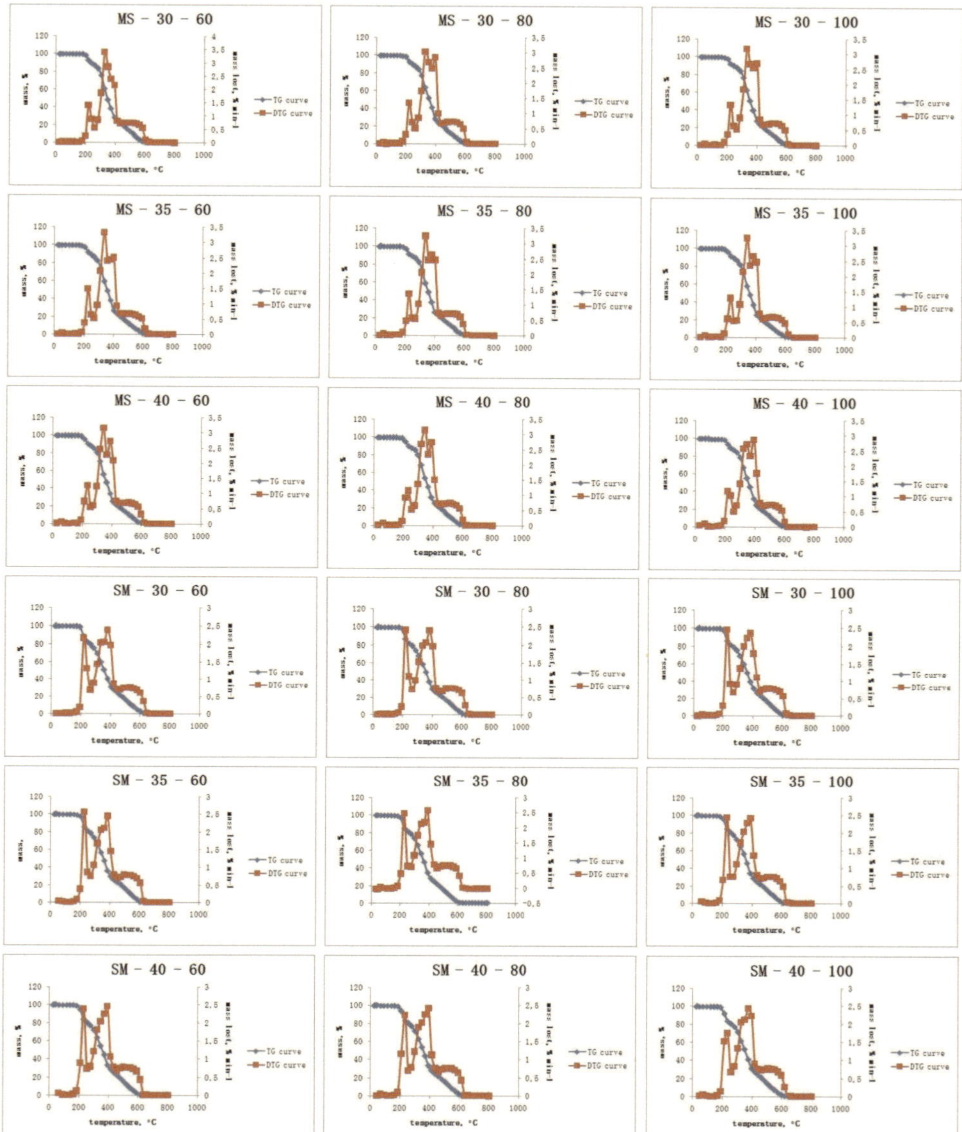

Figure 6.
TG curves and DTG curves of spreads with 70% sucrose/30% maltitol and 70% maltitol/30% sucrose.

score from 4.0–4.44, 4.33–4.61 and 4.11–4.5. Increasing the process parameters and maltitol concentration, the scores for flavor, taste, and chewiness are greater. Unique and complex flavor of the chocolate is one of the most important properties that have made it popular among the consumers [49]. Generally, the effect on sensory properties depends not only on the type of polyol but also on the polyol concentrations and process parameters.

4.7 Troubleshooting

Some troubleshooting can occur during the process production.

If the spread/mass is too viscous after refining/conching, the crystals of sugar alcohols are melted and release water. This problem could be solved by decreasing

Quality factor	Impact factor	Temperature (°C)								
		30			35			40		
Mixer speed rotation (o/min)		60	80	100	60	80	100	60	80	100
						MS				
External appearance	0.6	2.7	2.7	2.7	2.7	2.7	2.7	2.7	2.7	2.4
Texture	0.8	3.6	3.6	3.2	3.6	3.6	4	3.6	4	3.2
Chewiness	1	4.5	4.5	4	4.5	5	5	4	5	4
Flavor	0.6	2.4	2.4	2.4	2.4	2.4	2.4	2.4	2.4	2.4
Taste	1	4	4.5	4.5	4	4.5	4.5	4.5	4.5	4
Σ		17.2	17.7	16.8	17.2	18.2	18.6	17.2	18.6	16.0
Quality category		VG	E	VG	VG	E	E	VG	E	VG
Mixer speed rotation (o/min)						SM				
External appearance	0.6	2.1	2.1	2.1	2.1	2.1	2.1	2.4	2.4	2.4
Texture	0.8	2.8	2.8	2.8	2.8	2.8	2.8	2.8	3.6	2.8
Chewiness	1	4	4	4	4	4	4.5	4.5	4.5	4
Flavor	0.6	2.4	2.4	2.7	2.4	2.4	2.4	2.4	2.7	2.4
Taste	1	4	4.5	4.5	4	4	4.5	4	5	4.5
Σ		15.3	15.8	16.1	15.3	15.3	16.3	16.1	18.2	16.1
Quality category		VG	VG	VG	VG	VG	VG	VG	E	VG

Table 7.
Sensory characteristics of spreads with sucrose and maltitol.

Quality factor	Impact factor	Temperature (°C)								
		30			35			40		
Mixer speed rotation (o/min)		60	80	100	60	80	100	60	80	100
						S				
External appearance	0.6	2.4	1.8	1.5	1.5	1.5	1.5	1.5	1.5	2.1
Texture	0.8	3.2	2.8	2.4	2.8	2.8	2.4	2.8	2.8	3.6
Chewiness	1	4	4	4	4	4	4	4.5	4.5	4.5
Flavor	0.6	3	2.7	3	2.7	3	2.4	3	3	3
Taste	1	4	4	4.5	4.5	4.5	4.5	4.5	4.5	5
Σ		16.6	15.3	15.4	15.5	15.8	14.8	16.3	16.3	18.2
Quality category		VG	VG	VG	VG	VG	VG	VG	VG	E
Mixer speed rotation (o/min)						M				
External appearance	0.6	2.1	2.7	2.7	2.4	2.7	3.0	2.1	2.4	2.4
Texture	0.8	3.2	3.6	4.0	2.8	3.6	3.6	3.2	3.6	4
Chewiness	1	4	4	4	3.5	4	4.5	4	4.5	4.5
Flavor	0.6	2.4	2.7	2.4	2.4	3	3	2.7	2.7	2.7
Taste	1	4	4	4.5	4	5	5	5	5	5
Σ		15.7	17.0	17.6	15.1	18.3	19.1	17.0	18.2	18.6
Quality category		VG	VG	E	VG	E	E	VG	E	E

Table 8.
Sensory characteristics of spreads with 70% sucrose/30% maltitol and 30% sucrose/70% maltitol.

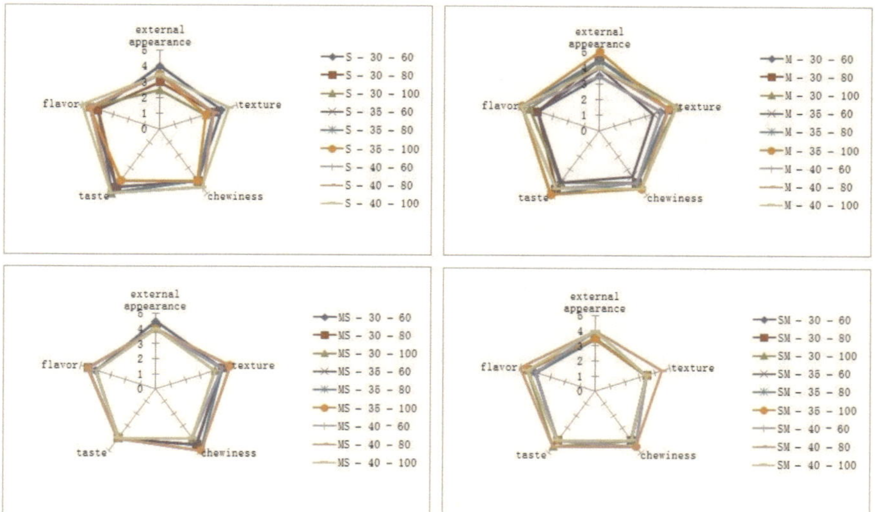

Figure 7.
QDA diagram of spreads.

the process temperature below the melting point of selected sugar alcohol or using the polyol with higher melting temperature and less hygroscopy. This problem could be solved by adding the fat too.

If the spread mass is grainy (sandiness), the moister content should be a problem. This problem could be solved by checking the storage conditions (low relative stability and temperature), polyol recrystallization (especially sorbitol), and polyol particle size and uniformity. The moisture variation (too high moisture) could form the spread mass to thin, as a consequence of hygroscopic polyols, inadequate solid content, and/or process parameters.

5. Conclusions

Consumption of sugar-enriched food has risen dramatically over the past few years. Sugar-enriched food contributes extra calories usually without nutritional values and has a negative effect to human health. Sugar alcohols (polyols) have drawn the significant attention of consumers and producers too. Since these polyols are contemporary, there is a need for enquiring sugar-free products, as a rapidly growing category.

With variable properties and functionalities of polyols, it is essential to think about perfect conditions for the polyol application. A sucrose-free spread with maltitol as a bulking agent was successfully developed. Such spread is compatible with traditional spread with sucrose because the sweetness of maltitol is close to sweetness of sucrose and no additional artificial sweeteners may be needed. The influence of maltitol on rheological, textural, thermal, and sensory properties is dependent on the present levels of maltitol and process parameters (temperature, mixer speed rotation).

The present study demonstrates that spread maltitol resulted in similar properties to spread with sucrose. It can be recommended as an adequate sugar substitute in spread formulations since sugar-free spread was accepted very well among panelists of different ages.

The findings of this study indicate that maltitol, as a sugar substitute bulking agents, has the potential as a pleasant food in the processing of diabetic and reduced calorie spread.

Future experiments will be focused on the optimization and determination of the sugar-free formulation recipe and the effects of bulk sweeteners and process parameters, based on physicochemical and sensory properties.

Acknowledgements

The author of this chapter is grateful to Prof. Dr. Biljana Pajin, a full-time professor of the Faculty of Technology Novi Sad, for her unselfish help and collegiality.

Author details

Marko Petković
Faculty of Agronomy, Čačak, Serbia

*Address all correspondence to: marko.pele.petkovic@gmail.com

References

[1] Bobroff L. Nutrition and diet. In: O'Neil KW, Peterson RL, editors. Optimal Aging. Sarasota, FL: Optimal Aging, LLC; 2004. pp. 627-641

[2] Petkovic M, Pajin B, Tomic J, Torbica A, Serse Z, Zaric D. Textural and sensory properties of spreads with sucrose and maltitol. Chemical Industry. 2012;**66**(6):385-394. DOI: 10.2298/HEMIND110902094P

[3] O' Brien Nabors L, editor. Alternative Sweeteners. 3rd ed. New York: World Wide Web; 2001. pp. 1-13, 31-39. ISBN: 0-8247-0437-1

[4] Wheeler ML, Pi-Sunyer X. Carbohydrate issues. Type and amount. Journal of the American Dietetic Association. 2008;**108**:34-39. DOI: 10.1016/j.jada.2008.01.024

[5] Fitch C, Keim KS. Position of the Academy of Nutrition and Dietetics: Use of nutritive and nonnutritive sweeteners. Journal of the Academy of Nutrition and Dietetics. 2012;**112**(5): 739-758. DOI: 10.1016/j.jand.2012.03.009

[6] Regulation EC no 1333/2008 of the European Parliament and of the Council of 16 December 2008 on food additives. Accessed: June 10, 2014. Available from: http://eurlex.europa.eu/LexUriServ/LexUriServ.do?uri=OJ:L:2008:354:0016:0033:en:PDF

[7] Grabitske HA, Slavin JL. Perspectives in practice low-digestible carbohydrates in practice. Journal of the American Dietetic Association. 2008;**108**(10): 1677-1681. DOI: 10.1016/j.jada.2008.07.010

[8] EFSA. Scientific opinion on the substantiation of health claims related to the sugar replacers xylitol, D-tagatose, xylitol, sorbitol, mannitol, maltitol, lactitol, isomalt, erythritol, D-tagatose, isomaltulose, sucralose and polydextrose and maintenance of tooth mineralization by decreasing tooth demineralization. EFSA Journal. 2011; **9**(4):2076. Available from: http://www.efsa.europa.eu/en/efsajournal/doc/2076.pdf [Accessed: June 10, 2014]

[9] Chattopadhyay S, Raychaudhuri U, Chakraborty R. Artificial sweeteners—a review. Journal of Food Science and Technology. 2014;**51**(4):611-621. DOI: 10.1007/s13197-011-0571-1

[10] Grembecka M. Sugar alcohols—their role in the modern world of sweeteners: A review. European Food Research and Technology. 2015;**241**(1): 1-14. DOI: 10.1007/s00217-015-2437-7

[11] Ladret M, Le Bot Y, Nesvadba S, Ostermann E, Ribadeau-Dumas G. Polyols: their properties and applications in sugar-free chewing gum. In: Fritz D, editor. Formulation and Production of Chewing and Bubble Gum. Ltd. London: Kennedy's Publications Ltd; 2006. pp. 133-155. ISBN: 0-904725-10-3

[12] Perko R, DeCock P. Reduced-calorie bulk sweeteners. In: Mitchell H, editor. Sweeteners and Sugar Alternatives in Food Technology. Oxford: Blackwell Publishing; 2006. pp. 149-325 ISBN-13: 978-14051-3434-7

[13] Nelson AL. Sweeteners: Alternative. Minnesota: Eagan Press; 2000. pp. 31-59. ISDN: 1-891127-11-X

[14] Shankar P, Ahuja S, Sriram K. Non-nutritive sweeteners: Review and update. Nutrition. 2013;**29**(11–12): 1293-1299. DOI: 10.1016/j.nut.2013.03.024

[15] Xiao J, Li X, Min X, Sakaguchi E. Mannitol improves absorption and retention of calcium and magnesium in

growing rats. Nutrition. 2013;**29**(1): 325-331. DOI: 10.1016/j.nut.2012.06.010

[16] Petkovic M, Pajin B, Tomic J. Effects of temperature and mixer speed rotation on rheological properties of spreads with maltitol. Journal of Food Process Engineering. 2013;**36**:633-644. DOI: 10.1111/jfpe.12027

[17] Petkovic M. The effects of producing parameters on physical properties, thermal characteristics and quality of spreads with maltitol [thesis]. Novi Sad: Faculty of Technology University of Novi Sad; 2012

[18] Petkovic M, Seres Z, Pajin B, Banjac V. Thermal decomposition of maltitol spreads. Journal of Thermal Analysis and Calorimetry. 2014;**117**(1): 277-284. DOI: 10.1007/s10973-014-3659-9

[19] Oba S, Toker OS, Palabyik I, Konar N, Goktas H, Cukur Y, et al. Rheological and melting properties of sucrose-free dark chocolate. International Journal of Food Properties. 2017. DOI: 10.1080/10942912.2017.1362652

[20] Afoakwa EO, Paterson A, Fowler M. Factors influencing rheological and textural qualities in chocolate—A review. Trends in Food Science and Technology. 2007;**18**(6):290-298. DOI: 10.1016/j.tifs.2007.02.002

[21] Fernandes VA, Müller AJ, Sandoval AJ. Thermal, structural and rheological characteristics of dark chocolate with different compositions. Journal of Food Engineering. 2013;**116**(1):97-108. DOI: 10.1016/j.jfoodeng.2012.12.002

[22] Toker OS, Sagdic O, Şener D, Konar N, Zorlucan T, Dağlıoğlu O. The influence of particle size on some physicochemical, rheological and melting properties and volatile compound profile of compound chocolate and cocolin samples. European Food Research and Technology. 2016;**242**(8):1253-1266. DOI: 10.1007/s00217-015-2629-1

[23] O'Brien L. A Sweet Alternative to Sugar Reduction. Calorie Control Council; 2014. www.culinology.org/d/do/237

[24] Ghosh S, Sudha ML. A review on polyols: new frontiers for health-based bakery products. International Journal of Food Sciences and Nutrition. 2012; **63**(3):372-379. DOI: 10.3109/09637486.2011.627846

[25] Ronda F, Gomez M, Blanco CA, Caballero PA. Effects of polyols and nondigestible oligosaccharides on the quality of sugar-free sponge cakes. Food Chemistry. 2004;**90**(4):549-555. DOI: 10.1016/j.foodchem.2004.05.023

[26] Jang SJ, Kim HW, Hwang KE, Song DH, Kim YJ, Ham YK, et al. Effects of replacing sucrose with various sugar alcohols on quality properties of semi-dried Jerky. Korean Journal of Food Science of Animal Resources. 2015; **35**(5):622-629. DOI: 10.5851/kosfa.2015.35.5.622

[27] Kim GD, Jung EY, Seo HW, Joo ST, Yang HS. Textural and sensory properties of pork jerky adjusted with tenderizers or humectant. Korean Journal of Food Science of Animal Resources. 2010;**30**(6):930-937. DOI: 10.5851/kosfa.2010.30.6.930

[28] Raitore S, Peterson DG. Effects of polyol type and particle size on flavor release in chewing gum. Food Chemistry. 2018;**253**:293-299. DOI: 10.1016/j.foodchem.2018.01.123

[29] Lawson P. Erythritol. Sweeteners. UK: Blackwell Publishing and Leatherhead Publishing; 2007. pp. 153-166

[30] Kearsley MW, Maltitol BN. In: O'Brien Nabors L, editor. Alternative

Sweeteners. Boca Raton: CRC Press, Taylor & Francis Group. 2012

[31] Kearsley MW, Deis RC. Maltitol powder. In: O'Donnell K, Kearsley MW, editors. Sweeteners and Sugar Alternatives in Food Technology. West Sussex: Wiley-Blackwell; 2012

[32] Respondek F, Hilpipre C, Chauveau P, Cazaubiel M, Gendre D, Maudet C, et al. Digestive tolerance and postprandial glycaemic and insulinaemic responses after consumption of dairy desserts containing maltitol and fructo-oligosaccharides in adults. European Journal of Clinical Nutrition. 2014; **68**(5):575-580. DOI: 10.1038/ejcn.2014.30

[33] Grabitske HA, Slavin JL. Gastrointestinal effects of low-digestible carbohydrates. Critical Reviews in Food Science and Nutrition. 2009;**49**(4): 327-360. DOI: 10.1080/10408390802067126

[34] Sokmen A, Mint GG. Influence of some bulk sweeteners on rheological properties of chocolate. LWT: Food and Science Technology. 2006;**6**:1053-1058. DOI: 10.1016/j.lwt.2006.03.002

[35] FAO-WHO Food Standards. Codex Alimentarius. 2014. GSFA Online. Maltitol. Available from: http://www.codexalimentarius.net/gsfaonline/additives/details.html?id=159 [Accessed: May 7, 2014]

[36] Evrendilek GA. Sugar alcohols (polyols). In: Varzakas T, Labropoulos A, Anestis S, editors. Sweeteners: Nutritional Aspects, Applications, and Production Technology. Boca Raton: CRC Press; 2012

[37] Livesey G. Glycaemic responses and toleration. In: O'Donnell K, Kearsley MW, editors. Sweeteners and Sugar Alternatives in Food Technology. West Sussex: Wiley-Blackwell; 2012

[38] Khud SE, Jackson LS, Fu TJ, Williams KM. Effects of processing on the recovery of food allergens from a model dark chocolate matrix. Food Chemistry. 2015;**168**:580-587. DOI: 10.1016/j.foodchem.2014.07.084

[39] Lechter A. Tempering chocolate—A review. The Manufacturing Confectioner. 2009;**89**(1):53-58

[40] Norton JE, Fryer PJ. Investigation of changes in formulation and processing parameters on the physical properties of cocoa butter emulsions. Journal of Food Engineering. 2012;**113**(2):329-336. DOI: 10.1016/j.jfoodeng.2012.05.025

[41] Glicerina V, Balestra F, Dalla Rosa M, Romani S. Effect of manufacturing process on the microstructural and rheological properties of milk chocolate. Journal of Food Engineering. 2015;**145**: 45-50. DOI: 10.1016/j.jfoodeng.2014.06.039

[42] Abbasi S, Farzanmehr H. Optimization of the formulation of prebiotic milk chocolate based on rheological properties. Food Technology and Biotechnology. 2009;**47**(4): 396-403. ISSN 1330-9862

[43] Afaokwa EO, Paterson A, Fowler M, Vieira J. Modelling tempering behavior of dark chocolates from varying particle size distribution and fat content using response surface methodology. Innovative Food Science and Emerging Technologies. 2008;**9**:527-533. DOI: 10.1016/j.ifset.2008.02.002

[44] Afoakwa E, Paterson A, Fowler M, Vieira J. Microstructure and mechanical properties related to particle size distribution and composition in dark chocolate. International Journal of Food Science and Technology. 2009;**44**(1): 111-119. DOI: 10.1111/j.1365-2621.2007.01677.x

[45] Pirouzian HR, Damirchi SA. Rheological properties of sugarfree milk

chocolate: Comparative study and optimization. Food technology and economy, engineering and physical properties. Czech Journal of Food Science. 2017;**35, 5**:440-448. DOI: 10.17221/231/2016-CJFS

[46] Konar N. Influence of conching temperature and some bulk sweeteners on physical and rheological properties of prebiotic milk chocolate containing inulin. European Food Research and Technology. 2013;**236**(1):135-143. DOI: 10.1007/s00217-012-1873-x

[47] Steffe FJ. Casson plastic viscosity and yield value: what they are and what they mean to the confectioner. In: Rheological Methods in Food Process Engineering. 2nd ed. Michigan: Freeman Press; 1996. pp. 94-250. ISBN: 0-9632036-1-4

[48] Svanberg L, Ahrne L, Loren N, Windhab E. Effect of sugar, cocoa particles and lecithin on cocoa butter crystallization in seeded and non-seeded chocolate model systems. Journal of Food Engineering. 2011;**104**(1):70-80. DOI: 10.1016/j.jfoodeng.2010.09.023

[49] Owusu M. Influence of raw material and processing on aroma in chocolate [PhD thesis]. Faculty of Life Science, University of Copenhagen; 2010

Chapter 4

Kinetic Approach to Multilayer Sorption: Equations of Isotherm and Applications

Francisco S. Pantuso, María L. Gómez Castro, Claudia C. Larregain, Ethel Coscarello and Roberto J. Aguerre

Abstract

An analytical isotherm equation that describes the multilayer adsorption on fractal surfaces with adsorbate-adsorbate interactions (measured in terms of free energy) different from that of bulk liquid was developed. Assuming mathematical functionalities for the variation of the free energy, it is possible to evaluate the influence of the adsorbate-adsorbate interactions on the adsorption capacity of solids of high degree of surface irregularity. For those surfaces with relatively low degree of irregularity, it results that the free energy variation with the layer number in the multilayer region affects considerably the sorption capacity of the adsorbent, even for water activities lower than those corresponding to the monolayer moisture content. The energy interactions between adjacent adsorbate layers become less important as the fractal dimension of the adsorbent increases. For a fractal surface, the growing of the multilayer seems to mainly controlled by the degree of surface roughness characteristic of microporous adsorbents, where the volume and pore dimension are the true limitants to the sorption capacity. The isotherm equations obtained were tested fitting published experimental equilibrium data of various water vapor-biopolymer systems.

Keywords: isotherm, roughness, multilayer, fractal, adsorption, free energy

1. Introduction

It is well known that the knowledge and understanding of water adsorption isotherms is of great importance in food technology. This knowledge is highly important for the design and optimization of drying equipment, packaging of foods, prediction of quality, and stability during storage.

In order to describe the overall sorption over the whole region of relative pressures of water, an isotherm for multilayer sorption must be used.

In 1938, Brunauer, Emmett, and Teller (BET) [1] extended Langmuir's monolayer theory [2–5] to multilayer adsorption. The BET equation derived was applied to a wide variety of gases on surfaces as well as to the sorption of water vapor by food materials [6–8].

But, the simple BET equation gives a good agreement with experimental data only at relative pressures lower than 0.35 of adsorbate. A great number of researchers have been analyzed this worrying fact, and numerous modifications have been proposed to the BET model to amend this problem [9–11].

Among these, the three parameters of GAB equation [12–15], introduce a modification to the BET sorption model. The GAB model is basically similar to BET ones in its assumptions. These authors propose that the state of the adsorbed molecules beyond the first layer is the same but different from that in the liquid state. This equation describes satisfactorily the sorption of water vapor in foods up to water activities of 0.8–0.9 [16–19]. The main advantage of the GAB equation is that its parameters have physical meaning. This equation has been adopted by West European Food Researchers [20].

For water activities higher than 0.8–0.9, most of the food materials show values of moisture content larger than that predicted by the GAB model. This flaw indicates that state of the adsorbed molecules beyond the first layer introduced by the GAB model is limited to a certain number of sorption layers. Then turn up as plausible to assume a third stage for the water molecules in the outer zone with true liquid-like properties, as postulated by the original BET model.

A three-zone model for the structure of water near water/solid interfaces was proposed by Drost-Hansen [21]; in this model, beyond the monolayer, a zone of ordered molecular structures of water is expected to exist adjacent to a surface, the ordering extending into the bulk liquid. This is a transition region over which one structure decays into another. At sufficiently large distances from the surface, bulk water structure exists.

The BET model and its modifications were developed for an energetically homogeneous flat surface without lateral interaction and are not suitable for highly rough surfaces [22].

This roughness plays a significant role in the determination of the adsorption characteristics [23–25], since the shape of the adsorbent surface influences the accessibility of the adsorbate to the active adsorption sites. In this chapter, their fractal dimension will characterize the roughness of the adsorbing surfaces. In addition, taking into account the model of the three zones, the derivation of an equation is presented for BET type multilayer isotherms on rough surfaces. This equation takes into account the influence of the adsorbate-adsorbent interaction of all the adsorbed layers.

It is shown that under certain conditions, this equation is reduced to the known classical forms. The capacity of the different isothermal equations to adjust the equilibrium moisture in the food is analyzed.

2. Mathematical model

Brunauer, Emmett, and Teller proposed an adsorption surface divided into *n* segments, having 1, 2, 3, ..., i number of layers of adsorbed molecules. According to this model, adsorption and desorption occur at the top of these segments. So, the equilibrium between the uncovered surface s_o and the first layer s_1^* is:

$$a_1 \frac{P}{P^{\ddagger}} s_0 = b_1 s_1^* \exp\left(-\frac{E_1}{RT}\right) \tag{1}$$

where a_1 and b_1 are adsorption and desorption coefficients, the same meaning as in BET theory, E_1 is the heat of adsorption of the first layer, R is the gas constant, T is the temperature, and P is the vapor pressure of adsorbate. Between any successive layers, the equilibrium can be expressed as:

DOI: http://dx.doi.org/10.5772/intechopen.82669

$$a_i \frac{P}{P^{\ddagger}} s^*_{i-1} = b_i s^*_i \exp\left(-\frac{E_i}{RT}\right) \tag{2}$$

Being s^*_{i-1} and s^*_i the surfaces at the top of the respective i−1 and i layers. Considering that R ln (b_i/a_i) is the sorption entropy of the i-layer, Eq. (2) can be written in a more convenient form:

$$\frac{P}{P^{\ddagger}} = \frac{s^*_i}{s^*_{i-1}} \exp\left(-\frac{\Delta G_i}{RT}\right) \tag{3}$$

where ΔG_i is the sorption free energy of the i-layer. This development differs from the classical BET model in that ΔG_i for all layers above the first is not considered equal to free energy of bulk liquid adsorbate, ΔG_L. Assuming that the free energy of sorption for the i-layer differs from the free energy of bulk liquid by a certain amount, it can be written in general that:

$$\Delta G_i = \Delta G_L + \Delta G^e_i \tag{4}$$

where ΔG^e_i differentiates the state of the adsorbed molecules from that of the molecules in the pure liquid. Substituting Eq. (4) in Eq. (3), it results:

$$s^*_i = \omega_i s^*_{i-1} \tag{5}$$

where

$$\omega_i = \frac{P}{P^{\ddagger}} \exp\left(\frac{\Delta G_L}{RT}\right) \exp\left(\frac{\Delta G^e_i}{RT}\right) \tag{6}$$

Defining

$$h_i = \exp\left(\frac{\Delta G^e_i}{RT}\right) \tag{7}$$

and given that $P_0 = P^{\ddagger} \exp(-\Delta G_L/RT)$, it results:

$$\omega_i = \frac{P}{P_0} h_i = x h_i \tag{8}$$

being $x = P/P_0$. The fraction of surface occupied by 1st, 2nd, ..., ith layer follows the relation:

$$s^*_i = s^*_1 \prod_{j=2}^{i} \omega_j \tag{9}$$

Combining Eqs. (8) and (9), we have

$$s^*_i = s^*_1 \prod_{j=2}^{i} (x h_j) \tag{10}$$

As $s^*_1 = \omega_1 s_0$, it results:

$$s^*_i = h_1 s_0 x^i \prod_{j=2}^{i} h_j = C\, s_0 x^i \prod_{j=2}^{i} h_j \tag{11}$$

where $C = h_1 = \exp(\Delta G^e_1/RT)$ is the constant C of BET theory.

Given that the adsorbate molecules, considered as spheres, when adsorbed osculate the surface, for a fractal surface the relationship between the surfaces at the top, s_i^*, and the bottom, s_i, for a molecular stack of i-layers is [26]:

$$s_i^* = s_i(2i\text{–}1)^{2\text{–}D} \tag{12}$$

where D is the fractal dimension; when Eq. (12) is substituted in Eq. (11), it gives:

$$s_i = C\, s_0 x^i (2i - 1)^{D-2} \prod_{j=2}^{i} h_j \tag{13}$$

According to the BET theory, the monolayer capacity, N_m, is:

$$N_m = \frac{1}{\sigma}\sum_{i=0}^{\infty} s_i = \frac{s_0}{\sigma}\left[1 + Cx + \sum_{i=2}^{\infty} C\, x^i (2i - 1)^{D-2} \prod_{j=2}^{i} h_j\right] \tag{14}$$

where σ is the cross-sectional area of water molecule. The total amount of adsorbent in a given layer n is:

$$N_n^t = \frac{(2n\text{–}1)^{2\text{–}D}}{\sigma}\sum_{i=n}^{\infty} s_i = \frac{C\, s_0}{\sigma}(2n\text{–}1)^{2\text{–}D}\sum_{i=n}^{\infty}(2i - 1)^{D-2} x^i \prod_{j=2}^{i} h_j \tag{15}$$

The total number of molecules, N, that form the adsorbed film is:

$$N = \frac{1}{\sigma}\left[s_1 + \sum_{i=2}^{\infty} s_i \sum_{k=1}^{i}(2k\text{–}1)^{2\text{–}D}\right] = \frac{Cs_0}{\sigma}\left[x + \sum_{i=2}^{\infty} x^i (2i - 1)^{D\text{–}2} \prod_{j=2}^{i} h_j \sum_{k=1}^{i}(2k\text{–}1)^{2\text{–}D}\right] \tag{16}$$

but, from Eq. (15), it is also:

$$N = \sum_{i=1}^{\infty} N_i^t = \frac{C\, s_0}{\sigma}\left[x + \sum_{i=2}^{\infty}(2i\text{–}1)^{2\text{–}D}\sum_{j=i}^{\infty} x^j (2j - 1)^{D-2}\prod_{k=2}^{j} h_k\right] \tag{17}$$

Finally, combining Eqs. (14) and (16), the following general equation for sorption isotherms is found:

$$\frac{N}{N_m} = \frac{C\left[x + \sum_{i=2}^{\infty} x^i (2i\text{–}1)^{D\text{–}2} \cdot \prod_{j=2}^{i} h_i \sum_{k=1}^{i}(2k\text{–}1)^{2\text{–}D}\right]}{\left[1 + Cx + \sum_{i=2}^{\infty} C\, x^i (2i\text{–}1)^{D\text{–}2}\prod_{j=2}^{i} h_j\right]} \tag{18}$$

But combining Eqs. (14) and (17), other equivalent form of the equation for sorption isotherms is reached:

$$\frac{N}{N_m} = \frac{C\left[x + \sum_{i=2}^{\infty}(2i\text{–}1)^{2\text{–}D}\sum_{j=i}^{\infty} x^j (2j - 1)^{D-2}\prod_{k=2}^{j} h_k\right]}{\left[1 + Cx + \sum_{i=2}^{\infty} C\, x^i (2i - 1)^{D-2}\prod_{j=2}^{i} h_j\right]} \tag{19}$$

Eq. (18) is therefore the isotherm equation for multilayer adsorption on fractal surfaces that takes into account the variation of the free energy of adsorption with successive layers.

2.1 Applications of the model to smooth surfaces

For a nonfractal surface (D = 2), Eqs. (18) and (19) reduce, respectively, to:

$$\frac{N}{N_m} = \frac{C\left[x + \sum_{i=2}^{\infty} i x^i \cdot \prod_{j=2}^{i} h_i\right]}{\left[1 + Cx + C\sum_{i=2}^{\infty} x^i \prod_{j=2}^{i} h_j\right]} \quad (20)$$

$$\frac{N}{N_m} = \frac{C\left[x + \sum_{i=2}^{\infty}\sum_{j=i}^{\infty} x^j \prod_{k=2}^{j} h_k\right]}{\left[1 + Cx + C\sum_{i=2}^{\infty} x^i \prod_{j=2}^{i} h_j\right]} \quad (21)$$

It is interesting to comment that for h_i = 1 and i ≥ 2 (free energy of the multilayer equal to the free energy of bulk water), Eq. (20) reduces to BET equation (see Eq. (22) in **Table 1**).

Even more, if for the second and higher layers the free energy of the adsorbate differs from that of pure liquid in a constant amount (h = k), Eq. (18) reduces to GAB equation (see Eq. (23) in **Table 1**).

Assuming that the adsorbate properties approach to the pure liquid as i increases ($\lim_{i\to\infty} \Delta G_i^e = 0$), $h_2 > h_3 > \ldots > 1$ or $h_2 < h_3 < \ldots < 1$, depending on arrangement of the adsorbate in the multilayer region (see Eqs. (24) and (25) in **Table 1**).

Eqs. (22) and (23) are frequently used in bibliography and tested for different adsorbate/adsorbent systems [29–31].

The ability of Eq. (24) to fit experimental data of water sorption on different food products is presented in **Table 2**.

In **Table 3**, the fitting test corresponding to Eq. (25) can be seen.

Eq. (25) gives a good agreement using data of amilaceous materials, nuts, and meats, whereas Eq. (24) shows a good fitting with fruits, some vegetables, and milk products.

2.1.1 Limited sorption

If the number of the adsorbed layers cannot exceed some finite number *n*, then for D = 2, h_i = 1, and i ≥ 2, from Eq. (18)

$$\frac{N}{N_m} = \frac{Cx}{(1-x)} \frac{\left[1-(n+1)x^n + nx^{n+1}\right]}{\left[1-x + Cx - Cx^{n+1}\right]} \quad (28)$$

also obtained by Brunauer et al. [1].

But from Eq. (19), D = 2, h_i = 1, and i ≥ 2

h_i	Equation	References
1	$\frac{N}{N_m} = \frac{Cx}{(1-x)[1-x+Cx]}$ (22)	[1]
k	$\frac{N}{N_m} = \frac{Ckx}{(1-kx)[1-kx+Ckx]}$ (23)	[27]
$\frac{i}{i-1}$	$\frac{N}{N_m} = \frac{Cx(1+x)}{(1-x)\left((1-x)^2+Cx\right)}$ (24)	[28]
$\frac{i-1}{i}$	$\frac{N}{N_m} = \frac{Cx}{(1-x)[1-C\cdot\ln(1-x)]}$ (25)	[28]

Table 1.
Equations derived from Eq. (20).

Material	Temperature	C	N_m%, d.b.	E%*	References
Milk products					
Edam cheese (a)	25°C	27.6	2.1	4.7	[32]
Emmental cheese (a)	25°C	77.7	1.9	10.3	
Yoghurt (a)	25°C	25.2	2.8	2.2	
Fruits					
Apple (d)	20°C	60.3	4.6	7.1	[33]
Apricots (a)	30°C	1.7	3.9	10.4	
Banana (a)	25°C	1.5	4.8	18.1	[32]
Figs (a)	30°C	1.7	4.9	5.8	[33]
Pear (a)	25°C	6.1	5.2	6.2	[32]
Pineapple (a)	25°C	4.7	4.7	25	
Plums (a)	30°C	1.9	4.9	10.8	[33]
Sultana raisins (a)	30°C	48.4	4.5	8.3	
Vegetables					
Carrots (s)	37°C	2.9	4.5	5.8	[34]
Onion, tender (a)	25°C	64.7	4.1	5	[32]
Radish (a)	25°C	9.7	4.6	8.4	
Spinach (s)	37°C	64.3	3.2	1.2	[34]
Sugar beet (d)	25°C	14.3	4.3	6.4	[35]

(a), adsorption; (d), desorption, (s), sorption.
*$E\% = 100 \cdot \sum_n |N_p - N_e|/N_p$ p : predicted e : experimental

Table 2.
Food sorption isotherms fitted with Eq. (24).

$$\frac{N}{N_m} = \frac{Cx}{(1-x)}\frac{(1-x^n)}{(1-x+Cx)} \tag{29}$$

known as Pickett [10] or Rounsley [50] isotherm equation.

2.2 The fractal isotherm

If the surface of the adsorbent behaves like a fractal with 2 < D < 3 and assuming that the free energy distribution in the adsorbed film is the same like in BET theory (h_i = 1), the following fractal isotherm equations can be obtained from Eqs. (18) and (19), respectively:

$$\frac{N}{N_m} = \frac{C\left[x + \sum_{i=2}^{\infty} x^i (2i-1)^{D-2} \sum_{j=1}^{i} (2j-1)^{2-D}\right]}{\left[1 + Cx + C\sum_{i=2}^{\infty} x^i (2i-1)^{D-2}\right]} \tag{30}$$

$$\frac{N}{N_m} = \frac{C\left[x + \sum_{i=2}^{\infty} (2i-1)^{2-D} \sum_{j=i}^{\infty} x^j (2j-1)^{D-2}\right]}{\left[1 + Cx + C\sum_{i=2}^{\infty} x^i (2i-1)^{D-2}\right]} \tag{31}$$

To illustrate the effect of roughness on the shape of the isotherms, in **Figure 1**, the influence of D values, for C = 20, can be seen.

Material	Temperature	C	N_m%, d.b.	E%*	References
Starchy foods					
Barley (d)	25°C	16.6	8.4	7.5	[36]
Corn shelled (d)	25°C	21.6	8.5	7.5	[37]
Flour (a)	30°C	57.7	6.8	9.1	[38]
Hard red winter (d)	25°C	23.1	9.3	5.4	[39]
Native manioc starch (d)	25°C	22.4	9.5	6.8	[37]
Native potato starch (a)	25°C	11.9	11.5	3.1	
Oats (d)	25°C	16.2	8	8.2	[37]
Rough rice (d)	60°C	8.2	5.7	3.2	[40]
Rye (d)	25°C	26.4	8.3	7	[37]
Sorghum (a)	38°C	19.0	8.4	6.1	[41]
Starch (a)	30°C	9.6	9.4	1.6	[38]
Tapioc starch (d)	25°C	19.9	10.8	8.2	[37]
Wheat, durum (d)	25°C	22.1	9.1	5	[39]
Nuts and oilseeds					
Moroccan sweet almonds (a)	25°C	6.5	3.7	2.2	[42]
Para nut (a)	25°C	9.4	2.7	4.7	[32]
Peas, dried (a)	25°C	75.1	8.3	3.3	[42]
Pecan nut (a)	25°C	6.8	2.9	4.7	[32]
Rapeseed, guile (d)	25°C	10.7	4.6	3.2	[43]
Rapeseed, tower (a)	25°C	8.4	5	2.3	
Soybean seed (s)	15°C	5.6	8	5.4	[44]
Meats					
Beef, minced (a)	30°C	4.4	9.2	5.4	[45]
Beef, raw/minced (s)	10 °C	2.3	14.3	8.4	[46]
Cod, freeze dried/unsalted (a)	25°C	5.6	12.2	5.8	[47]
Fish flour (a)	25°C	4.2	6.8	5.3	[48]
Mullet roe, unsalted (a)	25°C	5.3	4.8	1.4	[49]
Mullet, white muscle (a)	25°C	6.1	10.1	3.7	

(a), adsorption; (d), desorption, (s), sorption.
$^{*}E\% = 100 \cdot \sum_n |N_p - N_e|/N_p \quad p: predicted \quad e: experimental$

Table 3.
Food sorption isotherms fitted with Eq. (25).

Table 4 illustrates the usefulness of Eq. (30) on starchy materials.

In all the modeled materials, it is found that the value of the parameter D is approximately 2.8. This indicates that the tested products show a high roughness.

2.3 A four parameters equation

Assuming that the total surface area available for sorption is formed by two types of surfaces or regions: (a) a region representing a fraction, α, of the total adsorbing

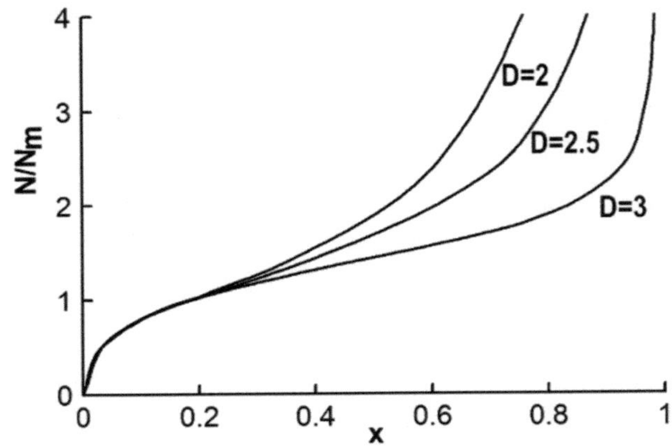

Figure 1.
Shape of the isotherm for different values of D (C = 20).

Material	Temperature	C	N_m%, d.b.	D	E%*	Source
Starches						
Wheat (a)	30°C	9.1	8.1	2.8	4.3	[38]
Corn (d)	25°C	12.5	9.8	2.9	1.0	[37]
Potato (a)	25°C	4.0	9.5	2.8	3.3	[51]
Tapioca (d)	25°C	14.3	10.2	2.9	2.4	[37]
Native manioc (a)	25°C	15.8	9.3	2.8	1.7	[37]
Cereals						
Rough rice (d)	25°C	10.7	9.0	2.9	2.6	[52]
Rough rice (d)	40°C	8.3	8.9	2.9	1.3	[53]
Sorghum (d)	20°C	16.4	9.6	2.9	0.9	
Sorghum (d)	50°C	6.9	8.8	2.9	1.1	
Wheat durum (d)	25°C	12.6	8.7	2.8	0.9	[39]
Corn, continental (d)	20°C	12.5	9.5	2.9	1.0	[53]
Corn, continental (d)	50°C	4.0	7.9	2.8	1.9	

(a), adsorption; (d), desorption; (s) sorption.
*$E\% = 100 \cdot \sum_n |N_p - N_e|/N_p$ *p : predicted e : experimental*

Table 4.
Food sorption isotherms fitted with Eq. (30).

surface that only adsorbs a limited number of adsorbate layers, that is, internal surface such as pores; (b) the remainder fraction, $(1 - \alpha)$, of the total adsorbing surface, where unlimited sorption may occur, that is, external surface and macropores where large values of n are required to fill them. From Eq. (29), it can be written [54]:

$$\frac{N}{N_m} = \frac{Cx(1 - \alpha x^n)}{(1 - x)(1 - x + Cx)} \tag{32}$$

In **Figure 2**, data of champignon mushroom and its fit using Eqs. (32) and (29) for different temperatures are presented.

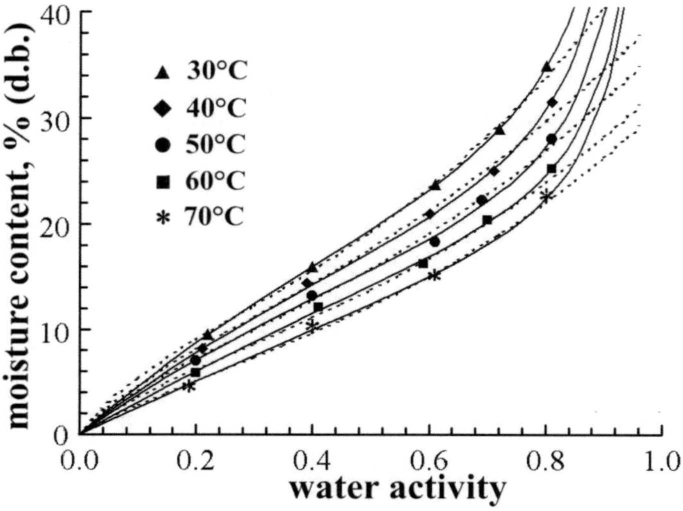

Figure 2.
*Influence of temperature on desorption isotherms of champignon mushroom (*Agaricus bisporus*). Solid line, Eq. (32); dotted line, Pickett Eq. (29).*

In **Table 5**, the fitting of moisture sorption on different products using Eq. (32) can be seen.

Eq. (32) gives a good agreement with experimental data. The inclusion of the alpha parameter allows modifying the amplitude of the isotherm plateau.

Material	T°C	C	*Nm*%, d.b.	α	n	E%[*]	References
Champignon mushroom (a) (*Agaricus bisporus*)	30	2.42	20.28	0.8478	1.41	0.19	[55]
	40	2.36	18.52	0.8556	1.37	1.49	
	50	2.61	15.59	0.8588	1.50	1.91	
	60	1.79	17.80	0.9037	1.26	1.78	
	70	1.64	15.68	0.8945	1.30	2.71	
Casein (a)	25	12.04	5.74	0.9864	3.72	0.97	[51]
Casein (d)	25	9.72	7.96	0.8923	2.34	0.77	
Coffee (a)	20	2.21	3.23	0.8784	14.94	3.62	
Dextrin (a)	10	12.02	13.01	0.8752	0.99	0.98	
Potato starch (a)	20	8.57	8.03	0.9230	3.47	1.21	
Anis (a)	25	15.03	4.32	0.9900	9.42	1.15	[32]
Avocado (a)	25	10.55	3.52	0.9900	10.13	3.43	
Banana (a)	25	0.41	18.65	0.6432	3.87	9.06	
Cardamom (a)	25	25.30	6.02	0.6734	2.64	0.67	
Celery (a)	25	5.39	7.06	0.8406	9.09	3.43	
Chamomile (a)	25	16.78	6.07	0.5010	4.67	0.94	
Emmenthal (a)	25	9.68	3.42	0.9604	13.25	1.66	
Cinnamon (a)	25	20.18	6.26	0.9999	3.63	0.19	
Clove (a)	25	29.72	4.25	0.6638	3.08	1.45	
Coriander (a)	25	10.95	5.87	0.7814	2.21	0.84	
Eggplant (a)	25	6.56	7.76	0.4396	7.96	4.00	

Material	T°C	C	Nm%, d.b.	α	n	E%*	References
Fennel (a)	25	0.33	70.06	0.9777	0.22	7.60	
Forelle (a)	45	6.63	5.12	0.9185	12.79	4.86	
Ginger (a)	25	15.23	7.30	0.8049	2.29	1.11	
Horseradish (a)	25	17.02	6.80	0.6437	5.00	1.31	
Huhn (a)	45	6.89	5.39	0.9253	10.93	1.99	
Joghurt (a)	25	5.18	5.15	0.9421	19.53	8.07	
Laurel (a)	25	17.58	4.38	0.8125	5.95	2.37	
Lentils (a)	25	17.67	6.95	0.8701	3.63	0.76	
Marjoram (a)	25	20.24	4.94	0.4006	2.91	2.37	
Mint (a)	25	12.69	7.42	0.6493	1.90	0.88	
Nutmeg (a)	25	26.93	4.57	0.8170	2.82	0.86	
Para nut (a)	25	26.85	1.81	0.7100	4.10	0.90	
Pears (a)	25	1.64	12.31	0.9200	7.96	2.65	
Pecan nut (a)	25	13.60	1.95	0.9869	4.90	1.38	
Pineapple (a)	25	0.46	26.24	0.9688	2.63	3.95	
Radish (a)	25	1.86	10.71	0.9901	7.98	13.20	
Savory (a)	25	28.36	6.61	0.8497	3.44	1.97	
Thyme (a)	25	23.39	4.77	0.6729	4.43	1.44	
Rice, rough (d)	40	3.12	11.30	0.9148	1.45	0.56	[56]
	50	2.60	11.02	0.9264	1.34	1.68	
	60	1.94	11.20	0.9520	1.39	1.67	
	70	1.56	11.72	0.9508	1.28	2.10	
	80	1.40	10.94	0.9544	1.45	2.17	
Meat, raw minced (s)	10	11.41	6.48	0.9807	10.71	2.42	[46]
	30	7.82	5.99	0.9711	10.20	4.91	
	50	13.67	4.78	0.9746	12.80	1.54	
Lard (s)	25	13.96	0.36	0.6626	4.10	1.07	
Mullet roe, unsalted (a)	25	3.14	7.37	0.7957	0.65	0.73	[49]
Mullet, white muscle (a)	25	9.46	7.12	0.9681	4.86	1.73	
Cod, unsalted (a)	25	13.02	7.78	0.6336	3.67	4.79	[47]

(a), adsorption; (d), desorption; (s), sorption.
*$E\% = 100 \cdot \sum_n |N_p - N_e|/N_p$ p : *predicted* e : *experimental*

Table 5.
Food sorption isotherms fitted with Eq. (32).

The modeling of the sigmoid isotherms is facilitated, typical form found in the adsorption of water in food products.

3. Conclusions

In the framework of the BET model, a general isotherm equation was obtained that includes the roughness of the adsorbent surface and characterizes the transition region between the monolayer and the outer zone where the adsorbate has the properties of the bulk liquid through a free energy excess that differentiates the adsorbed phase from the bulk liquid.

This general equation, depending on the simplifications assumed, gives the classical BET and GAB equations. But, taking into account an asymptotic reduction of the free energy excess, for a flat surface, two different equations were obtained. One of them appears useful to model starchy materials, extending the isotherm plateau, and the other, successfully model fruit isotherms, reducing the isotherm plateau.

Considering only the roughness, and assigning bulk liquid properties at all layers beyond the first, an equation that includes the fractal dimension is obtained.

This fractal dimension can vary from 2 to 3. The rising of its value result in an outspread the isotherm plateau. Particularly, for highly rough surfaces, the multilayer growing is limited by geometrical restrictions. In this case, the magnitude of the interactions practically has no effect on the shape of the isotherm.

It results from the present analysis that modifications of the BET model based only on the three-zone model or geometric considerations conduct to similar results.

So, Eqs. (25), (30), and (32) predict lower sorption capacity with the increment of water activity, giving better agreements with experimental isotherms than the classical BET equation.

This fact forewarns that the fractal dimension in the model could be affected from unsuitable accounting for the adsorbate-adsorbent interactions.

Acknowledgements

The authors gratefully acknowledge the Morón University for providing financial support.

Author details

Francisco S. Pantuso, María L. Gómez Castro, Claudia C. Larregain, Ethel Coscarello and Roberto J. Aguerre*
Agroalimentary Research Laboratory, FAyCA, Morón University, Morón, Buenos Aires, Argentina

*Address all correspondence to: rjaguerre@gmail.com

References

[1] Brunauer S, Emmett PH, Teller J. Adsorption of gases in multimolecular layers. Journal of the American Chemical Society. 1938;**68**:309-319

[2] Saadi R, Saadi Z, Fazaeli R, Fard NE. Monolayer and multilayer adsorption isotherm models for sorption from aqueous media. Korean Journal of Chemical Engineering. 2015;**32**(5):787-799. DOI: 10.1007/s11814-015-0053-7

[3] Zhang P, Wang L. Extended Langmuir equation for correlating multilayer adsorption. Separation and Purification Technology. 2010;**70**: 367-371

[4] Patiha, Firdaus M,Wahyuningsih S, Nugrahaningtyas KD, Hidayat Y. Derivation and constants determination of the Freundlich and (fractal) Langmuir adsorption isotherms from kinetics. IOP Conference Series: Materials Science and Engineering. 2018;**333**:1-9

[5] Thomas S, Schaetzel P. Multilayer adsorption equilibrium model for gas adsorption on solids. Adsorption. 2013; **19**:121-129

[6] Chen S-W, Guo B-L, Wang Y-L, Li Y, Song L-J. Study on sorption of U(VI) onto ordered mesoporous silicas. Journal of Radioanalytical and Nuclear Chemistry. 2013;**295**:1435-1442. DOI: 10.1007/s10967-012-1998-1

[7] Cherik D, Louhab K. A kinetics, isotherms and thermodynamic study of diclofenac adsorption using activated carbon prepared from olive stones. Journal of Dispersion Science and Technology. 2017;**39**(6):814-825. DOI: 10.1080/01932691.2017.1395346

[8] Charoen R, Tipkanon S, Savedboworn W. Sorption isotherm study of preserved wild mangosteen (kra-thon yhee) with replacing humectants. International Food Research Journal. 2018;**25**(3):1258-1265

[9] Hill TL. Theory of Physical Adsorption Advances in Catalysis. Vol. 4. New York: Academic Press; 1952. pp. 211-257

[10] Pickett G. Modification of the Brunauer–Emmett–Teller theory of multimolecular adsorption. Journal of the American Chemical Society. 1945; **67**:1958-1962

[11] Bashiri H, Orouji H. A new isotherm for multilayer gas adsorption on heterogeneous solid surfaces. Theoretical Chemistry Accounts. 2015; **134**:1594-1600. DOI: 10.1007/s00214-014-1594-2

[12] Anderson RH. Modification of the BET equation. Journal of the American Chemical Society. 1946;**68**:689-691

[13] de Boer JH. The Dynamical Character of Adsorption. 2nd ed. Oxford: Clarendon Press; 1968. 217 p

[14] Guggenheim EA. Application of Statistical Mechanics, Chap. 11. Oxford: Clarendon Press; 1966

[15] Arthur E, Tuller M, Moldrup P, Greve MH, Knadel M, de Jonge LW. Applicability of the Guggenheim–Anderson–Boer water vapour sorption model for estimation of soil specific surface area. European Journal of Soil Science. 2018;**69**:245-255. DOI: 10.1111/ejss.12524

[16] van den Berg C. Vapour sorption equilibrium and other water-starch interactions; A physico-chemical approach [PhD thesis]. Wageningen: Agricultural University; 1981

[17] Knani S, Aouaini F, Bahloul N, Khalfaoui M, Hachicha MA,

Ben Laminea A, et al. Modeling of adsorption isotherms of water vapor on Tunisian olive leaves using statistical mechanical formulation. Physica A. 2014;**400**:57-70

[18] Staudt PB, Tessaro IC, Marczak LDF, Soares RDP, Cardozo NSM. A new method for predicting sorption isotherms at different temperatures: Extension to the GAB model. Journal of Food Engineering. 2013;**118**:247-255

[19] Zou L, Gong L, Xu P, Feng G, Liu H. Modified GAB model for correlating multilayer adsorption equilibrium data. Separation and Purification Technology. 2016;**161**:38-43

[20] Van den Berg C, Bruin S. Water activity and its relation in food systems: Theoretical aspects. In: Rockland LB, Stewart GF, editors. Water Activity: Influences on Food Quality. London: Academic Press; 1981

[21] Drost-Hansen W. Structure of water near solid interfaces. Industrial and Engineering Chemistry. 1969;**61**(11): 10-47

[22] Fripiat JJ. Flow and porosimetry. In: Avnir D, editor. The Fractal Approach to Heterogeneous Chemistry. Surfaces, Colloids, Polymers. Chichester/New York/Brisbane/Toronto/Singapore: John Wiley & Sons; 1990

[23] Vajda P, Felinger A. Multilayer adsorption on fractal surfaces. Journal of Chromatography A. 2014;**1324**: 121-127

[24] Bashiri H, Shajari A. Theoretical study of fractal-like kinetics of adsorption. Adsorption Science & Technology. 2014;**32**:623-634

[25] Xu YF, Matsuoka H, Sun DA. Swelling characteristics of fractal-textured bentonite and its mixtures. Applied Clay Science. 2003;**22**:197-209

[26] Aguerre RJ, Viollaz PE, Suarez C. A fractal isotherm for multilayer adsorption in foods. Journal of Food Engineering. 1996;**30**:227-238

[27] Blahovec J, Yanniotis S. 'Gab' generalised equation as a basis for sorption spectral analysis. Czech Journal of Food Sciences. 2010;**28**(5): 345-354

[28] Aguerre RJ, Suarez C, Viollaz PE. New BET type multilayer sorption isotherms. Part II: Modeling water sorption in foods. Lebensmittel-Wissenschaft und-Technologie/FST. 1989;**22**:192-195. ISSN: (0023-6438)

[29] Sawhney IK, Sarkar BC, Patil GR. Moisture sorption characteristics of dried acid casein from buffalo skim milk. LWT- Food Science and Technology. 2011;**44**:502-510

[30] Donnarummab D, Tomaiuolo G, Caserta S, Gizaw Y, Guido S. Water evaporation from porous media by dynamic vapor sorption. Colloids and Surfaces A: Physicochemical and Engineering Aspects. 2015;**480**:159-164

[31] Aouaini F, Knani S, Ben Yahia M, Bahloul N, Kechaou N, Ben Lamine A. Application of statistical physics on the modeling of water vapor desorption isotherms. Drying Technology: An International Journal. 2014;**32**(16): 1905-1922. DOI: 10.1080/07373937.2014.924131

[32] Wolf W, Spiess WEL, Jung G. Die Wasser dampf sorptions isothermen eihiger in der llteratur bislang wehig beruck sichtigter. Lebensmittel-Wissenschaft und Technologie. 1973; **6**(3):94-96

[33] Maraulis ZB, Tsami E, Marinos-Kouris D. Application of the GAB model to the moisture sorption isotherms for dried fruits. Journal of Food Engineering. 1988;7:63-78

[34] Makover B, Dehority GL. Equilibrium moisture content of dehydrated vegetables. Industrial and Engineering Chemistry. 1943;**35**(2): 193-197

[35] Iglesias HA. Isotermas de sorción de agua en remolacha azucarera y análisis del fenómenode sorción de agua en alimentos [Tesis]. Buenos Aires University, F.C.E.yN.; 1975

[36] Coleman DA, Fellows HC. Hygroscopic moisture of cereal grains and flaxseed exposed to different relative humidities. Cereal Chemistry. 1955;**2**:275-287

[37] Hellman NN, Melvin EH. Surface area of starch and its role in water sorption. Journal of the American Chemical Society. 1950;**72**:5186-5188

[38] Bushuk W, Winkler CA. Sorption of water vapor on wheat flour, starch and gluten. Cereal Chemistry. 1957;**34**:73-76

[39] Day DL, Nelson GL. Desorption isotherms for wheat. Transactions of ASAE. 1965;**8**:293-297

[40] Aguerre RJ, Suarez C, Viollaz P. Moisture desorption isotherms of rough rice. Journal of Food Technology. 1983; **18**:345-351

[41] Dunstan ER, Chung DS, Hodges TO. Adsorption and desorption characteristics of grain sorghum. Transactions of ASAE. 1973;**16**(4): 667-670

[42] Pixton SW, Henderson S. Moisture relations of dried peas, shelled almonds and lupins. Journal of Stored Products Research. 1979;**15**(2):59-63

[43] Pixton SW, Warburton S. The moisture content/equilibrium relative humidity relationship and oil composition of rapeseed. Journal of Stored Products Research. 1977;**13**:77-81

[44] Gane R. The water content of the seeds of peas, soya beans, linseed, grass, onion and carrot as a function of temperature and humidity of the atmosphere. Joumal of Agricultural Science. 1948;**38**:81-83

[45] Iglesias HA, Chirife J. Effect of fat content on the water sorption isotherm of air dried minced beef. Lebensmittel-Wissenschaft und Technologie. 1977;**10**: 151-153

[46] Motarjemi Y. A study of some physical properties of water in foodstuffs. Water activity, water binding and water diffusivity in minced meat products [PhD thesis]. Lund, Sweden: Lund University; 1988

[47] Doe PE, Hashmi R, Pouler RG, Olley J. Isohalic sorption isotherms. Journal of Food Technology. 1982;**17**:125-134

[48] Labuza TP, Kaanane A, Chen JY. Effect of temperature on the moisture sorption isotherms and water activity shift of two dehydrated foods. Journal of Food Science. 1985;**50**:385-391

[49] Chau KV, Heinis JJ, Perez M. Sorption isotherms and drying rates for mullet fillet and roe. Journal of Food Science. 1982;**47**:1318-1328

[50] Rounsley RR. Multimolecular adsorption equation. AIChE Journal. 1961;**7**(2):308-311

[51] Bizot H. Using the GAB model to construct sorption isotherms. In: Jowitt R, Escher F, Hallstrom B, Meffert HFT, Spiess WEL, Vos G, editors. Physical Properties of Foods. Vol. 11. London/ New York: Applied Science Publishers; 1983

[52] Zuritz CA, Singh RP, Moini SM, Henderson SM. Desorption isotherms of rough rice from 10 to 40°C. Transactions of ASAE. 1979;**22**:433-440

[53] Falabella MC, Aguerre RJ, Suarez C. Modelling non-isothermal sorption equilibrium data of cereal grains. Lebensmittel-Wissenschaft und Technologie. 1992;**25**:286-288

[54] Pantuso FS, Tolaba MP, Aguerre RJ. A BET approach to multilayer adsorption in swelling products. Journal of Food Commerce. 2014;**122**:68-71

[55] Shivhare US, Arora S, Ahmed J, Raghavan GSV. Moisture adsorption isotherms for mushroom. Lebensmittel-Wissenschaft und Technologie. 2004;**37**: 133-137

[56] Iguaz A, Vírseda P. Moisture desorption isotherms of rough rice at high temperatures. Journal of Food Engineering. 2007;**79**:794-802

Chapter 5

Risk Management of Egg and Egg Products: Advanced Methods Applied

Poliana Mendes de Souza, Regiane de Melo,
Miriam Aparecida de Aguilar Santos, Fabiana Regina Lima
and Kássia Héllen Vieira

Abstract

Generally, foods are thermally processed to destroy the vegetative microorganisms for food preservation. However, only thermal treatment triggers many unwanted biochemical reactions, which leads to undesirable sensorial and nutritional effects. Therefore, a number of nontraditional preservation techniques are being developed to satisfy consumer demand. Ensuring food safety and at the same time meeting such demands for retention of nutrition and quality attributes has resulted in increased interest in emerging preservation techniques. The techniques are mainly focused on liquid foods and/or vegetable matrix, a lack of data is observed concerning animal food products. On this way, this chapter discusses about the alternative technologies developed and implemented considering sanitation and preservation of eggs.

Keywords: shell eggs, sanitization, pasteurization food safety, microbiological contamination, alternative technologies

1. Introduction

The chicken egg is as considered one of the nature's most complete foods because of its high nutritional value. It is composed of a variety of nutrients, vitamins, minerals, fatty acids, and protein, which makes it one of the most important foods in human nutrition. These nutrients are efficiently absorbed and essential for the proper functioning of the human body. In addition, it has low cost and high availability in most countries, which makes it possible to increase the consumption of a food of high nutritional value by the low-income population [1].

The world consumption of eggs increases each year with a consequent increase in production. In 2014, the world production of eggs was around 1.275 trillion of units. China is the main producer (36%), followed by the United States (7.9%), India (6.0%), Mexico (4.0%), Brazil (3.5%), and Japan (3%). It is noteworthy that these countries are among the world's top 10 chicken egg-producing nations [2].

Eggs are composed of approximately 65% water, 12% protein, 11% lipid, and 12% ash; it also has low carbohydrate content and provides only 72 calories [3]. In addition, it is a source of water-soluble and fat-soluble vitamins such as retinol, tocopherol, ascorbic acid, riboflavin, pantothenic acid, and vitamin D and minerals such as calcium, iron, phosphorus, copper, and zinc [4]. Egg is considered as a food of high biological value, because it has all the amino acids required in human nutrition [5].

The egg has three main components: shell (11%), egg white (58%), and yolk (31%). In **Figure 1**, more details about the structures of the egg can be observed.

The egg white, corresponding to 58% of the whole egg, has a main component the water (about 88%), being low in fat and rich in protein. The albumen (egg white) is constituted of around 40 different types of proteins, which are responsible for the functional and antimicrobial characteristics of the egg white. The main proteins present in the egg are egg albumin (corresponding to 50% of the proteins), conalbumin, ovomucoid, lysozyme, ovomucin, avidin, and ovoglobulin [7]. In the egg white, there is also the presence of carbon dioxide, which makes it cloudy, but this substance tends to disappear in aged eggs, making it look more transparent than fresh eggs. The albumen has the ability to form foams; it is fundamental in the formulation of soufflés, meringues, and omelets, which is denatured at temperatures above 58°C [8].

The egg yolk is a central part that lies within the egg white and it is yellow in color, representing 31% of the egg, and contains three-quarters of the total value of calories [7]. Pigmentation of egg yolks may vary depending on the feed of birds; however, this variation has no influence on the quality or nutritional value of the egg. The majority of the egg nutrients are present in the yolk, which is composed mainly of lipids (34%), and proteins, such as lecithin and globular proteins [9]. The lecithin protein is responsible for the emulsification of products such as mayonnaise and Hollandaise sauce [10]. The egg yolk consists of about 50% water, and its denaturation occurs at temperatures above 62.5°C [7].

The term "egg products" refers to eggs that have been removed from their shells to undergo processing operations, whether they are breaking, filtering, blending, stabilizing, pasteurizing, cooling, freezing, drying, and/or packaging. This definition includes whole eggs, yolk, or egg white that have been processed, pasteurized, and can be found in liquid, frozen, or dehydrated form [6].

Eggs are consumed worldwide because they are highly versatile, allowing them to be used in various culinary preparations. It can be served alone or as an ingredient to provide improvement on texture, flavor, structure, moisture, and increase nutritional value. Eggs also have great importance in the food industry, due to their technological characteristics, such as incorporation of air, gelatinization, and emulsification, which are desirable in meringues, biscuits, bakery products, and meat products [9].

The production route starts on farms, where the eggs are taken to the warehouses for washing, classification, and packaging into packages made with expanded polystyrene or cellulose pulp with capacity for 12 or 30 eggs.

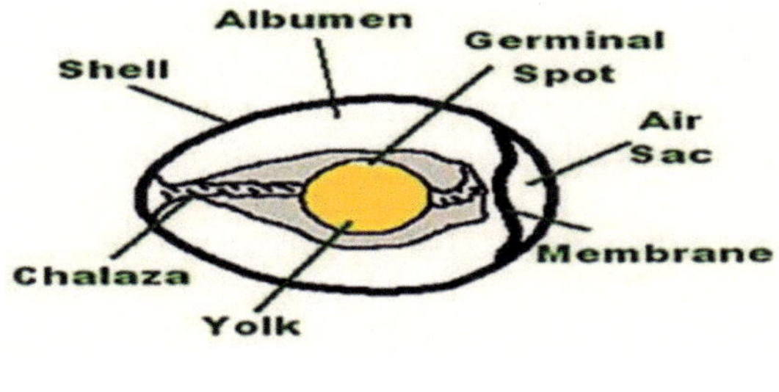

Figure 1.
Structure of the egg. Source: Souza [6].

Subsequently, they are packed in cardboard boxes and sent to the wholesale trade, in trucks, for retail resales [11].

Despite the nutritional value and functional properties of the egg, there are some problems resulting from its storage, which may interfere in quality. The fact that egg is a product rich in nutrients makes it conducive in the development of spoilage and pathogenic microorganisms [4]. Another important fact is that only 5% of the total chicken egg production in Brazil is destined for industrial processing. It is understood that the other 95% are intended for in natural consumption, where the eggs do not undergo quality control before being used in some preparation, as required in food industry [11]. Thus, the storage conditions of eggs, such as time and temperature are essential to ensure safety and quality, since they are packaged in their natural form and where there is a quality problem, it will be visible only to the consumer at the time of use.

In general, the eggs present little contamination at the moment of the posture, which usually occurs after the oviposition [12]. Eggs can be contaminated in contact with feces: by transovarian contamination (when the chicken's ovaries are infected) or by microorganism penetration through the pores and microscopic cracks in the shell, whether in the washing process, packaging transport, or storage [13]. The genera of bacteria that contribute most to the deterioration of eggs are *Pseudomonas*, *Acinetobacter*, *Proteus*, *Aeromonas*, *Alcaligenes*, *Escherichia*, *Micrococcus*, *Serratia*, *Enterobacter*, and *Flavobacterium*. Meanwhile, the pathogenic bacteria associated with eggs are *Salmonella*, *Staphylococcus*, *Campylobacter jejuni*, *Listeria monocytogenes,* and *Yersinia enterocolitica* [14].

In this way, eggs need to go through some treatment that prolongs their shelf life and also reduces the risk of contamination by foodborne microorganisms. These treatments might be thermal or nonthermal, and the latter are better known as alternative techniques to thermal treatment in eggs.

2. Heat treatment in eggs: pasteurization

In order to extend the shelf life of eggs and their products, and to reduce consumer risks related to foodborne pathogens such as *Salmonella*, it is necessary that these products go through pasteurization. Thereby, to avoid the deterioration of this food, the method with wide application is the thermal pasteurization [4].

The pasteurized egg is preferably used in the food industry, when compared to the product in nature. Besides maintaining flavor, color, nutritional value, and functional properties, this method presents operational advantages, such as reduction of losses and wastes, ease of measuring portions and less space for storage, and it saves time and labor [15].

Hot water, steam, microwave, radiofrequency (RF), and freeze-drying are some of the thermal methods used for the decontamination of eggs and egg products. Each of the methods use different range of temperature conditions [16].

Although decontamination methods using heat are efficient for microbial reduction, they can negatively affect the physical-chemical characteristics, nutritional content, and also sensorial properties, such as color and texture, making this type of food and its products less attractive to the consumer [4].

2.1 Hot water method

Currently, the use of hot water is the main method of pasteurization in whole eggs, but less than 1% of all shelled eggs are pasteurized [17]. Normally, this processing is carried out in an equipment known as a water bath (**Figure 2**). In this

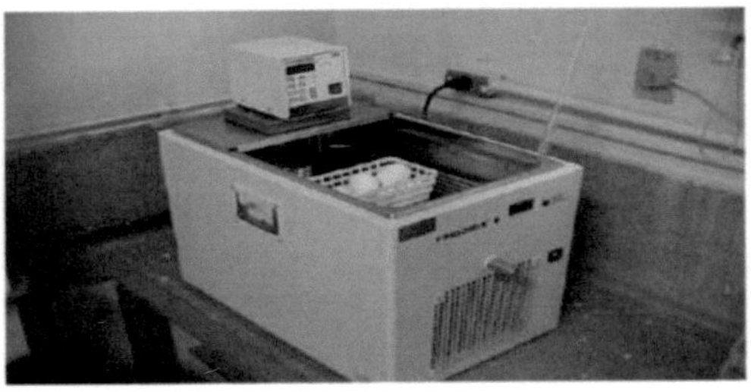

Figure 2.
Water bath equipment. Source: Pombo [11].

equipment, the water is heated to a certain temperature, the eggs are placed in the equipment and submitted to heating with defined time intervals [11].

Studies have shown that the load of inoculated *Salmonella typhimurium* cells was significantly reduced after the pasteurization process of shelled eggs in a circulating water bath at 57°C for 15 minutes [17].

Whole eggs submitted to pasteurization in water bath at 57°C for 20 minutes maintained their quality and showed a reduction of the microbial load [18].

In liquid eggs, the efficacy of the circulating water bath for *Salmonella enteritidis* inactivation at 65°C was verified in an interval of 0–7 minutes, already showing a reduction of the contamination at 3 minutes. This same process exerted less impact on the egg viscosity when compared to the high-pressure treatment, which has a positive effect on functionality and allows the use of liquid eggs in various products [19].

Decontamination by this method can also be accomplished by emerging the eggs (without cracking) in water at 95°C for 10 seconds. Studies have claimed that when liquid eggs are subjected to temperatures above 70°C for 1.5 seconds, there is a significant reduction of *Salmonella enteritidis* [20, 21]. However, despite the efficiency in egg decontamination, this method affected the egg quality, altering the texture, yolk membrane strength, albumin contents, and yolk characteristics [22].

However, egg washing may decrease or remove the cuticle layer that surrounds the eggshell (responsible for antimicrobial defense), increasing the probability of microbial invasion, reducing the quality and life of the washed eggs [23].

2.2 Steam method

Steam pasteurization may be a valuable alternative to egg surface decontamination, also in relation to the ban of the use of water by the European Union in eggs. However, further studies are made on the efficacy of decontamination of this technique on eggs. Among the available studies, there are studies that investigated the applicability of a steam gun treatment to pasteurization of the egg surface. They investigated the temperatures inside and outside the egg and identified that 180°C for 8 seconds as the best treatment corresponding to the surface temperature, the highest that can be achieved without detrimental changes to egg quality. Unfortunately, no microbiological investigation was performed [24].

Whole egg pasteurization can be completed by using steam generators with 60°C for 8 seconds, while eggs spin and swirl through the aid of mechanical engineering. Then, the eggs are treated with cold air through hot air generators (20–25°C) for 32 seconds. This treatment was effective in reducing *Salmonella enteritidis* and *Salmonella typhimurium* in egg shells and did not affect the egg quality [18, 24].

Whole egg pasteurization can be done using steam generators at a temperature lower than previously reported (heating at 60°C for 8 seconds) while the eggs roll through the aid of mechanical engineering. Then, the eggs are treated with cold air through hot air generators (20–25°C) for 32 seconds. This treatment was effective in reducing *Salmonella enteritidis* and *Salmonella typhimurium* in egg shells and did not affect egg quality. This method is further recommended for pasteurization of egg yolk, egg white, and whole egg liquid [17, 23].

After evaluation and comparison of the quality characteristics of eggs treated with steam and eggs in nature, after 28 days of storage at 20°C, it could be observed that the quality parameters (pH and color) were not different, indicating that the treatment of steam does not exert negative effects on the main quality characteristics of the egg. These parameters, along with the microbial results in experimentally inoculated eggs, suggest that the industrial application of steam treatment in eggs prior to the packaging is useful to achieve a reduction of approximately 90% of the population of *Salmonella enteritidis*, which naturally infects the surface of the eggs [24].

2.3 Microwave method

Microwave-assisted thermal method is a new thermal processing technology that provides rapid volumetric heating [25]. Electromagnetic waves are able to reduce *Salmonella enteritidis*, which is often found in shell eggs. The microwave frequency ranges from 300 MHz to 300 GHz, while the wavelength ranges from 1 mm to 1 m. In order to generate heat, the microwaves interact with dielectric materials and stir the molecules in an alternating electromagnetic field. Generally, foods have excellent microwave absorption capacity due to high water or carbon content, which can result in a faster temperature increase, thus requiring less time to inactivate the present microorganisms [26].

Microwave is an easy and affordable method to heat up food. However, the way absorbed energy is distributed depends on the shape, surface area, and food matrix, besides the type of equipment used. The eggs tend to burst when using this method to warm and sanitize, if the equipment exhibits high levels of energy. Therefore, while using this type of procedure, it is ideal to use low energy levels and slowly heating up the product [27, 28].

In **Figure 3**, it is possible to visualize a representative microwave scheme adapted at the laboratory level, which enables the measurement and control of the dielectric properties of the equipment from computer software.

Studies have shown that eggshell and eggshell membrane presented transparency to the microwave. The pasteurization of whole eggs, placed with the largest extremity face up, was achieved, when the shell was heated and the yolk reached the temperature of 61.1°C. A microwave oven with power 9 for 15 seconds showed efficiency in the reduction of previously inoculated *Salmonella* strains [18]. However, further investigations should be conducted regarding changes in egg rheology, viscosity, emulsifying property, and protein denaturation [16].

2.4 Radio-frequency heating

The radio-frequency band (RF) of the electromagnetic spectrum covers a wide range of high frequencies, typically in the kHz band ($3 \text{ kHz} < f \leq 1 \text{ MHz}$) or MHz band ($1 \text{ MHz} < f \leq 300 \text{ MHz}$) [30].

Unlike conventional systems, where thermal energy is transferred from a hot medium to a colder product resulting in large temperature gradients, radio-frequency heating involves the transfer of electromagnetic energy directly to the product, initiating heating due to friction, and interaction between molecules (heat is generated within the product). The RF heating is also known as high-frequency dielectric heating. During RF heating, the product to be heated forms a "dielectric" between two metal capacitor plates (electrodes) (**Figure 4**), which are alternately charged positively and negatively by a high electric current field [30].

RF heating is a promising application in food processing, due to the rapid and uniform spread of heat, better penetration, and low energy consumption. Researchers conducted on eggs using RF heating (10 MHz–3 GHz) using temperatures of 5–56°C indicated the eggshell and eggshell membrane are extremely transparent to this technology. The more transparent is the product investigated and pasteurized, and the more efficient is the decontamination [30, 32].

The immersion of the eggs in deionized water combined with the RF focused on the egg yolk and surface cooling showed a high security potential from the microbiological point of view [33]. The combination of RF (60 MHz) in water at 35°C for 3.5 minutes resulted in a temperature of 61°C inside the egg yolk. After that, the egg was again heated for another 20 min with water at 56.7°C. Performing this two-step process, with a total duration of 23.5 minutes, the *Escherichia coli* population significantly reduced (6.5 log); however, comparing with pasteurization only with

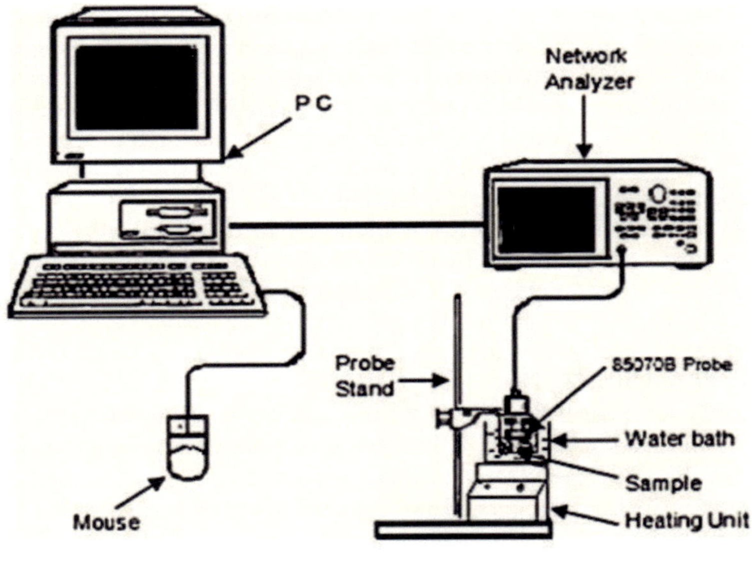

Figure 3.
Representative microwave scheme adapted at the laboratory level. Source: Dev et al. [29].

Figure 4.
Schematic representation of the RF heating process. Source: Marra et al. [31].

hot water, it took 60 minutes to reduce this microbial population by 6.6 log. The combination of the RF and hot water method was faster than the existing commercial process, using only hot water [34].

However, there is a disadvantage of RF, if it is not uniform, it can be observing the formation of coagulation rings around the egg air cell, thus damaging the decontamination process due to the impact on product quality [35].

2.5 Freeze-drying or cryodesiccation

Dehydration is a successful method of preserving eggs, it presents the advantages as follows: occupies less space in stock, provides ease of transport, good uniformity, easier use (ready-to-use product), and presents stable microbiological quality [36]. One of the main procedures used to dehydrate egg and turn it into powder is the freeze-drying or cryodesiccation method. This process consists of the rapid freezing (−50 to −60°C) of the liquid egg or part previously pasteurized and subsequent dehydration: the water contained in the product passes directly from the solid state to the vapor state by sublimation, under low temperature and vacuum conditions [36, 37].

A freeze-dryer or lyophilizer is used for the process of freeze-drying; it consists of a vacuum chamber, a heat source, a condenser, and a vacuum pump (**Figure 5**). The main function of the vacuum chamber (where the food is contained) is to resist the differences in pressure so the ice melting does not occur, and the pump helps to maintain this difference by removing noncondensable gases. The heat source is responsible for producing the energy that will evaporate the ice and is the type of source that determines the type of freeze-dryer used. The condenser retains moisture from the food and prevents its increase from inside the chamber and returning to the food [15].

The freeze-drying method applied on eggs occurs in three stages: (a) initial freezing of egg, (b) primary drying, in which the water is removed by sublimation that takes place under vacuum and the addition of heat and ends when the increase of the temperature of the egg is found in a value close to the environment or when it starts to defrost, and (c) secondary drying (also called desorption), which occurs after all ice has already been removed from the egg, but still retains an amount of liquid water (called tightly bound water), requiring a reduction of moisture to about 2–8%. For moisture reduction, the partially dried egg should be kept in the freeze-drying for about 2–6 hours and heated until its temperature equals that of the plate (20–60°C), maintaining the vacuum and evaporation of much of the wastewater [37, 38].

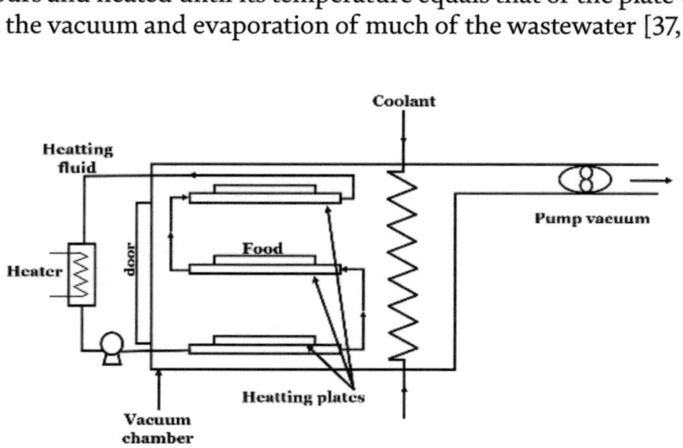

Figure 5.
Main components of the freeze-dryer. Source: Cunha et al. [15]—adaptado.

By going through the freeze-drying process, the egg and its products retain the sensory characteristics and the nutritional quality, because the temperature is not too high. In addition, they have an extended shelf life when packaged correctly. The volatile compounds are not absorbed by water vapor and are retained in the matrix of the products, allowing the retention of the egg aroma of about 80–100% [15, 39].

The main alteration that occurs in egg composition is the alteration in the quaternary and tertiary structures of the proteins. After water removal, changes occur in these structures due to exposure of the hydrophobic parts of the protein, previously protected inside the tertiary and quaternary structures, due to nonaffinity with water. The nutritional content and aroma of the eggs do not present significant changes. In addition, with the removal of water, preservation of egg powder is maintained, due to the low humidity, which reduces microbial proliferation [15, 37].

The main disadvantages of this method is that eggs may be susceptible to oxidation reactions (lipids, carotenoids, fat-soluble vitamins, and aromatic substances) if not packed into a vacuum, oxygen-impermeable, and opaque packaging [37, 39]. The freeze-drying process is time-consuming and may last up to 48 hours, depending on the batch size and the units to be processed, increasing the cost of the process. In addition, the freeze-dryer is a costly equipment [37, 40].

3. Alternative techniques for the thermal treatment in eggs

Considering that microorganisms are naturally present in any raw food, there are concerns regarding egg contamination. These products have the potential for contamination with bacteria from the animal's intestinal tract, feces, and the surrounding environment. In addition, eggs are an ideal growth medium for pathogenic bacteria that are dangerous to humans (*Salmonella, Escherichia, and Enterobacter*) [41].

Pasteurization techniques are used to prolong shelf life and maintain the quality of egg products. However, thermal techniques can have negative impact on the functional properties of this food, in the amount of nutrients, taste, and texture. Although heat processes used in egg pasteurization can ensure food safety by eliminating heat-sensitive pathogens, some heat-resistant microorganisms can survive the process, spoiling the product even under refrigerated conditions [42, 43]. In this way, new techniques are being developed and applied in the food industry.

New preservation technologies are an interesting option for producing high-quality food and extend its shelf life [44]. These technologies present a moderate impact on the sensory profile and quality attributes of the processed foods (such as flavor, color, aroma, and nutrients), giving food producers the opportunity to offer safe and high-quality food.

However, emerging techniques in the egg industry must be further studied, so they can be considered a successful processing and thus produce on a commercial scale. In this way, advantages and safety are provided not only to industry, but also to supermarkets and consumers [15, 45].

New food processing technologies include the use of physical factors to process and preserve food [46]. Among the new technologies, high hydrostatic pressure (HHP), pulsed electric fields (PEF), treatment with ozone, ultraviolet light (UV), and gamma radiation are nonthermal technologies with application in eggs and egg products.

3.1 High hydrostatic pressure

The application of HHP technology has attracted the interest of the food industry due to its microbial destruction capacity at very low or moderate temperatures, the preservation of bioactive nutrients, the improvement of the extraction of bioactive compounds, and the reduction of the allergenic potential of foods, such as eggs [47].

The HHP technology applies high pressures (usually in the range of 100 and 1000 MPa) with or without heat treatments, in order to eliminate different microorganisms and to guarantee the microbiological safety of the final product. This process is operated on a batch system, usually using water as a pressure transmission medium. The food products are packaged, loaded into the pressure vessel, and then pressurized by water [47, 48].

The HHP equipment is generally made of high-strength steel alloys, making it resistant to oxidation and rupture. HHP mainly is applied on batch equipment; however, semicontinuous systems are available. Generally, an HHP batch equipment consists of a pressure vessel (thick wall cylinder), two covers that close the pressure vessel, a yoke which controls the closing cover under the pressure condition, a pump and intensifier to create high pressure, and a process control system for loading and unloading the products [43]. An HHP batch system can be used for liquid and solid foods, while a semicontinuous HHP process can be used only for pumpable foods.

During the process, the food is packed, sealed, and loaded into a sample basket. The packaging shall consist of flexible materials, which will resist to pressurization. The sample baskets then enter the pressure vessel, which contains the pressure transmitting fluid. Water is usually used as pressure transmitting fluid on industrial scale equipment. The pump and the intensifier provide a desired pressure by compression of the pressure transmitting fluid. Thereafter, the product is maintained under the right time and pressure to achieve the desired treatment. At the end of the treatment, the vessel is depressurized and the product is unloaded from the sample basket [42, 43].

The application of heat combined with HHP can cause physical, chemical, or biological changes on the food product. These changes depend on the applied pressure, treatment time, and temperature and can include protein denaturation, changes in enzyme activities [42].

Many attempts have been made to verify if the HHP technique can be used as a substitute for thermal pasteurization, and to identifying the structural changes in the components of the egg as a result of the high pressure [44]. This technique has been evaluated as an alternative to methods already used for liquid eggs, and it has been verified that the processing conditions must be well studied, as this can cause a protein coagulation [45]. It has also been reported that pressure-induced protein denaturation may occur in eggs due to the entry of water into wells of the protein molecule [46]. However, HHP at a pressure between 200 and 350 MPa did not cause detectable protein denaturation in liquid eggs [47]. Other research has shown that HHP treatment on liquid eggs is a successful preservation opportunity. The application of 600 Mpa for a 2 minutes cycle in boiled eggs was able to extend the shelf life of these products during refrigeration [49].

HHP present important advantages for food processing, the fact that this technology does not produce deterioration of thermolabile nutrients (such as vitamins) and does not alter low molecular weight compounds, fundamentally those responsible for flavor and aroma. The high pressure does not favor the Maillard reaction or enzymatic browning; thus, it does not alter the natural flavor or color of the food [50, 51].

The application of HHP causes a number of changes in the morphology, cell membrane and biochemical reactions of microorganisms, and all these processes are related to microbial inactivation. In particular, the cell membrane is considered the main target for inactivation of microorganisms induced by pressure, and it is generally accepted that leakage of intracellular constituents across the permeabilized cell membrane is the most direct reason for cell death by high-pressure treatment [50, 52].

Research shows, in addition to the applied pressure level and treatment time, the critical parameters for microbial inactivation are pH, water activity (aw), and treatment temperature: (a) microorganisms become more susceptible to pressure at lower pH [53]; (b) water activity reduction exerts a protective effect on microorganisms against high-pressure treatments [54, 55]; and (c) thermal processing with temperatures above or below room temperature tends to increase the rate of inactivation of microorganisms [56].

The effect of microbiological inactivation on egg products by HHP was reported in a study that showed the low-pressure ranges are used to reduce the microbial load of liquid eggs by 3 log. The study concludes that increased pressure may increase the effectiveness of the treatment and thus lead to the processing of microbiologically safe products [57].

This technology has great potential for use in food processing, since it is efficient in the elimination of microorganisms, thus providing microbiological safety and increased shelf life, maintaining the nutritional and sensorial characteristics of foods [50].

3.2 Pulsed electric field

Pulsed electric field (PEF), or high-intensity electric field (HELP), is one of the nonthermal processing technologies of interest to scientists and the food industry; it is new and alternative method for preserving liquid foods. In addition, it is a promising alternative to traditional heat treatments, which presents good results, not only by enabling the destruction of microorganisms and the inactivation of enzymes, but also by maintaining the flavor, color, texture, vitamins and not only by enabling the destruction of microorganisms and the inactivation of enzymes, but also by maintaining the flavor, color, texture, vitamins, and functional thermolabile components [58].

Food processing by applying PEF involves subjecting the product to repeated electric fields (constituting the number of pulses) for short time intervals (microseconds) in order to inactivate enzymes and destroy microorganisms [59].

This method uses high voltage pulses on a treatment chamber containing food between two electrodes. The high electric intensity is acquired by accumulating a large amount of energy in a condenser, which supplies and discharges the energy in the form of pulses, for short periods of time, uniformly and with a minimum increase of temperature [60, 61]. **Figure 6** shows two types of treatment chambers used in the PEF process.

The PEF technology can be one of the most suitable methods for liquid food processing. In the last years, the technology received considerable attention from scientists, governments and interested industries as a potential technique to be fully expanded in the future year [45].

Research shows that PEF technology has been used successfully to pasteurize foods such as dairy products, a variety of fruit juices, liquid eggs, and creamy soups

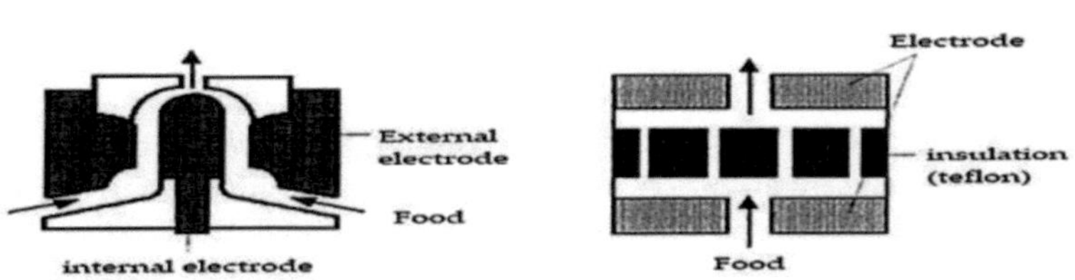

Figure 6.
Types of treatment chambers used in the PEF process. Source: Fani [62].

[63]. The application of PEF in the control of spoilage or pathogenic microorganisms in different egg products has been highlighted. It has been reported that this method effectively reduces the activity of numerous microorganisms in egg products [64].

Liquid egg is widely used by the food industry and other commercial food manufacturers due to the convenience, ease of handling, and longer shelf life compared to shell eggs. Egg is a polyfunctional ingredient because of its thickening, gelling, emulsifying, foaming, coloring, and flavoring attributes, which can be used to modify the organoleptic and technological properties of many food products. In addition, liquid egg products are also valuable because of their high-quality protein content and low cost [65, 66].

Although thermal treatments represent the most available pasteurization methods for liquid eggs, they can affect their functional properties and degrade the quality of the products. Thus, the application of PEF, as a nonthermal food processing technology, might be an alternative to conventional thermal preservation methods. Combined methods with PEF, such as homogenization, show a great potential to preserve the liquid egg with small modifications of its native color, viscosity, and foaming capacity [45].

Microbial inactivation by electrical pulses depends on several factors that are critical to treatment efficacy. These factors can be classified by process parameters (pulse intensity, treatment time, and temperature), product attributes (pH, ionic compounds, and conductivity), and characteristics of microorganisms (type, concentration, and growth stage) [67].

Gram-positive bacteria are more resistant to electrical pulsed treatment than Gram-negative bacteria; this factor may be due to the rigidity of the peptidoglycan layers present in its cell wall. Due to their larger size, fungi are more sensitive to this treatment than bacteria [68].

The exposure of a biological cell to a high-intensity pulsed electric field leads to a phenomenon of membrane permeabilization. This leads to pore formation, which is reversible if the electric field is below a certain critical value and for a short period of time. This phenomenon is called electroporation and is used in genetic engineering. However, overcoming certain values of field strength and processing time, this process becomes irreversible, results in loss of cellular material, and inactivation of the cell [69].

This method has advantages such as the treatment time, which is relatively short, provides a low-temperature pasteurization, is efficient in liquid products, maintains the sensorial characteristics of the product, and shows no evidence of toxicity. Thus, this technology can complement a heat treatment, or completely replace it. However, the PEF method is not indicated for solids or liquids containing air pockets [70].

It is not clear whether the food industry will fully accept PEF as a processing technology. However, the PEF is already being used industrially in some fruit juice industries in Europe. However, the PEF method has been used industrially in some fruit juice companies in Europe, presenting a number of applications growing over the year. Nevertheless, its potential to replace or complement conventional methods comes from research related to the use of PEF in all fields of food processing [71].

3.3 Ozone

In 2001, the Food and Drug Administration (FDA) approved the use of ozone (O_3), either in gas or liquid form, as a disinfectant to be applied in food processing and product stock. Since then, special attention has been given to the use of O_3 as a potent disinfectant to be used in a variety of environments, such as hospitals, candy factories, cheese maturation rooms, and poultry hatcheries. Besides its disinfectant performance, O_3 in gaseous form has the same properties for disinfecting eggs, fresh fruits, and vegetables. In liquid form, O_3 can be used to wash poultry and fish carcasses in order to reduce or even eliminate the microbial load [72, 73].

O_3 is a triatomic form of oxygen that has been gaining space on food processing due to its high sanitizing power and rapid degradation, leaving no waste on treated foods, and known as a highly reactive antimicrobial agent. Therefore, a hypothesis to increase the shelf life of eggs would be the exposure of it to O_3. In many research, O_3 has been shown to be very efficient in the inactivation of microorganisms that could degrade food [74–76].

Research has shown that the concentration of gaseous O_3 between 4 and 6 $mg.L^{-1}$ could be used to maintain the internal quality of the eggs and extend their shelf life. Concluding, gaseous O_3 present great potential as an emerging technology to maintain fresh egg quality and also extend shelf life during storage at room temperature [77].

Due to its high instability, O_3 must be produced at the place of disinfection and its use must be immediate because it decomposes rapidly into oxygen. O_3 is generated by the exposure of air, or other gas containing normal oxygen, to a high energy source. The production forms are by the method of electric discharge (corona discharge method), electrochemical methods, and UV radiation, all of them inspired by its natural formation in atmosphere. The electric discharge method is the most commercially used, even though it has low efficiency (2–10%) and high electricity consumption. The other methods are less cost effective, but the O_3 production by the UV method is less than by the electric discharge method because it only produces O_3 in a concentration of 0.1% by weight [78].

Besides the microbicidal effect, characteristics of lower toxicity and easy handling give the O_3 advantages of use. Added to these factors, its decomposition into nontoxic oxygen and rapid degradation characterize O_3 as a nonwaste-producing disinfectant [75].

The existence of several methods of measurement of O_3 in the environment is also one of the advantages of its use as a disinfectant. Physical, physical-chemical, and chemical methods are available in the market. Physical methods measure the direct absorption in the region of the electromagnetic spectrum UV, visible light, and infrared, while physical-chemical methods are dependent on effects such as heat or chemiluminescence caused by reactions. Chemical methods refer to the quantification of products when O_3 reacts with chemical reagents, such as potassium iodide (KI), the Indigo method the most recommended in this case [72].

However, O_3 cannot be considered universally beneficial to food, because in high concentrations, it can promote oxidative rancidity, so it can cause modification on taste and color of the food product. Changes in sensory or physical-chemical attributes depend on the chemical composition of the food, the O_3 dosage, and the treatment conditions [79].

3.4 Ultraviolet

UV radiation stands out as one of the few technologies that does not generate residue to the environment and is effective in reducing the microbial load when applied correctly. The application of the germicidal effects of UV radiation covers three categories: (a) inhibition of microorganisms on the surface, (b) destruction of microorganisms in the air, and (c) sterilization of liquids. Based on these effects, the use of UV light is widely used for sanitizing water and food processes [44].

Characteristics of practicality and low cost, combined with the advantage of not producing chemical residues, coproducts or radiation at the end of the process give UV-C an excellent alternative for disinfecting environments and products [45].

UV radiation comprises the portion of the electromagnetic spectrum ranging from about 100 to 400 nm. UV radiation is composed of different wavelengths greater than the X-ray (200 nm) and smaller than visible light (400 nm). The true UV radiation is actually invisible to the human eye, but its larger portion (around 400 nm) has a violet color, hence the name ultraviolet [44].

The wavelength of UV light can be divided into three bands: long-wave (UV-A, 400–320 nm), which occur in sunlight and has little germicidal value; medium-wave (UV-B, 320–280 nm), also found in sunlight and germicidal effect; while the short-wave (UV-C, 280–100 nm), has the greatest germicidal effects and does not occur naturally, which is produced by the electric energy conversion [46].

As mentioned, sunlight can be a source of UV rays; however, it is known that the range of solar UV-C radiation with greater germicidal potential is blocked by stratospheric ozone. Artificial sources of this radiation are obtained by mercury medium pressure and low pressure lamps, which produce energy in the germicidal region and which are electrically identical to fluorescent lamps, except for the absence of phosphorus cover. These lamps consist of an airtight silica or quartz tube (both UV transmitters), with the ends endowed with tungsten electrodes with a mixture of alkaline earth metal, which facilitates the formation of the electric arc inside the lamp. Inside the tube is introduced a small amount of mercury and an inert gas—usually argon. The voltage between the electrodes produces an excitation of mercury atoms, then when they return to a level of less energy, the excited molecules emit UV light. Low-pressure UV lamps—or monochromatic—emit 85–90% of radiation at the wavelength of 254 nm, with a higher germicidal effect. Thus, in the kinetic studies of UV disinfection, the mean intensity of the germicidal radiation considered is 254 nm. On medium-pressure lamps—or polychromatic lamps—the contributions of each radiation of different wavelength shall be taken into account when determining the dose [44, 47, 48]. **Figure** 7 shows a model of UV radiator.

There are two ways of applying UV light: pulsed and continuous light. The continuous mode is the conventional method; the light being emitted continuously without interruption. In pulsed UV-light mode, the UV-light is released as intermittent pulses using a capacitor, which allows to increase the energy intensity per pulse. Therefore, the pulsed mode is more effective for microbiological inactivation and the most used method [81].

The extent of UVC radiation to the microorganisms is conditional on the dose of radiation, which they can absorb. The dose required for destruction of the bacterial cell is relatively low and depends on the intensity and time of exposure. The impact of various obstacles can affect the optimum dose of UV light, because the light emitted by the germicidal lamp may not be absorbed by most of the microorganisms. Spores of microorganisms exhibit high UV resistance, and the sublethal dose may favor its growth rather than inhibiting it, so its use is important in environments and products with absence of organic matter and obstacles. In this way,

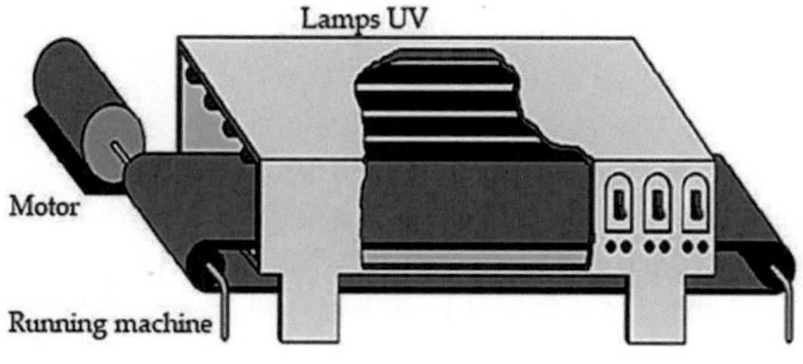

Figure 7.
UV radiation model. Source: Alexandre et al. [80].

microorganisms present on smooth and regular surfaces are more susceptible to the effects of UV light than those present on irregular surfaces [82–84].

Short-wave ultraviolet radiation (UV-C) has shown prospects for the pasteurization of liquid foods on appropriated reactors. Several studies have shown that the organoleptic properties of UV-treated liquid egg products are comparable to those untreated; therefore, those are excellent candidates for UV-C application. Regarding to the increase of the microbial load, research shows that the long-term microbial stability of liquid eggs was positively influenced by UV-C treatments and the shelf life was extended to 8 weeks in refrigerated storage. Thus, UV-C treatment is a promising technology to prolong the shelf life of liquid egg products [85, 86]. The time-temperature binomial is crucial to produce pasteurized eggs with high microbiological quality [87].

The biocidal effect occurs when UV-C radiation reaches the surface of the microorganism by overcoming the cell membrane and damaging its DNA genetic material. The DNA damage occurs through the formation of thymine dimers. The thymine dimer formation is the process of rupture of the nitrogen bases adenine and thymine (A-T), from the DNA. The rupture establishes a new chemical bond between two thymine, thus constituting the thymine dimers (T-T). The new binding prevents DNA replication and transcription, leading to the death of the microorganism [88–90].

Among the advantages of using UV-C radiation for food sterilization and disinfection are non-by-product production, does not alter sensory characteristics (taste, color, or odor), does not transmit radioactivity, it is a dry application process, and does not generate heat beyond the equipment and it is of low cost [91–93].

The main limitation of this technology involves the low degree of penetration that hinders the reach of the radiation by all the microbial load in the food. Thereby, it is more widely used in surface sterilization, for example in food packaging and in edible films. However, in liquid foods, the turbulent flow is recommended during processing [83, 94, 95].

3.5 Gamma radiation

Ionizing radiation, in the form of gamma rays, is obtained from isotopes or, commercially from X-rays and electrons, and it is applied in food preservation through microbial elimination or inhibition of biochemical changes. It has several advantages, such as low or no heat generation, low energy requirements, food preservation in a single operation, irradiation of packaged or frozen products, besides those it can cause changes in the nutritional value of food similar to other conservation methods [96].

Gamma rays are a type of electromagnetic radiation produced in nuclear decay processes. These are highly energetic due to their high frequency and consequently low wavelength. Generally, the frequency of the gamma rays is above 1019 Hz, which implies wavelengths below 10–12 m and energies above 0.1 MeV (the energy of the visible radiation ranges from 1 to 4 eV, about 50,000 times smaller) [97].

The irradiation is done in a special processing room or chamber for a certain time. The food is treated in an installation known as an irradiator. This equipment (**Figure 8**) consists of a cobalt-60 source installed in a bunker, which is an irradiation chamber whose walls are concrete shields, in the form of labyrinths. The radiation source, when the plant is not in operation, is stored in a pool (water well) with treated and demineralized water. The well is lined with a stainless steel coating, inside the shield. The food product to be irradiated is placed in containers and through a monorail are conducted into the irradiation chamber, where they receive

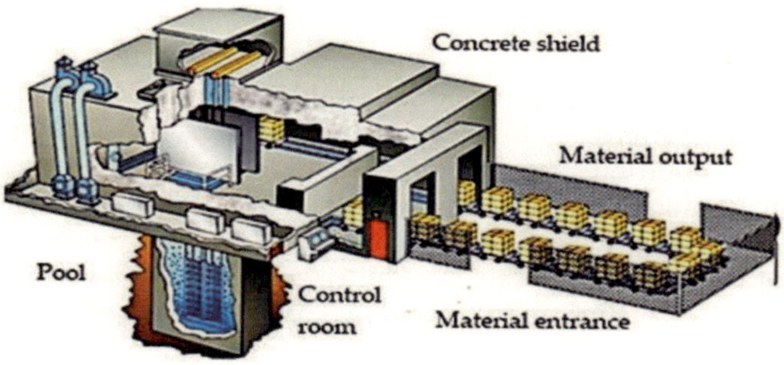

Figure 8.
Irradiation plant with cobalt-60 source. Source: Caldeira et al. [99]—adaptado.

the programmed dose of gamma radiation. Qualified operators electronically monitor the source of radiation and the treatment of the products from a console located in a room outside the irradiation chamber [98].

The gamma radiation sources commonly used on commercial plants are cobalt 60 and cesium 137. These isotope sources cannot be switched off, which is why they are kept in a water tank located below the processing area to allow an approximation of the machine operator. When the irradiator is in operation, the source is elevated and the packed food is transported by an automated conveyor through the irradiation field in a circular route that allows uniformity and efficiency of the process [96].

Regarding the inactivation of the microbial load, the efficiency of the treatment on the microorganisms depends on several factors: (a) the number of microorganisms: the higher the amount of microorganisms presents in the food, the higher the radiation dose required; (b) the food composition: microorganisms on rich media are more resistant than in buffer solution; (c) oxygen: the presence of oxygen makes the microorganisms less resistant to radiation; (d) state of matter: dehydrated or frozen cells are more resistant to radiation than in the normal state; (e) the condition of the microorganism: microorganisms in the lag phase are more resistant; and (f) the microorganism radioresistance: overall the more complex the DNA, the greater is the sensitivity of microorganisms to irradiation [96, 100].

The use of ionizing radiation is an alternative method in the reduction of pathogenic microorganism on eggs (such as *Salmonella* spp.) when the use of heat is impractical or undesirable for food preservation. Irradiation in an appropriate dose eliminates foodborne pathogens in frozen and unfrozen liquid eggs, powder egg white and egg yolk, fresh whole egg with intact peel, and cooked egg [101].

The advantages of the use of irradiation as a method of conservation over the other methods are: (a) time, since the irradiation can be applied in a few minutes; (b) the method does not leave residues in the food, because only the gamma rays come into contact with the food, without any risk of radioactive contamination; (c) it can be applied on a wide range of fruits; (d) it prevents food recontamination, since the product is already packed during the process; and (e) cold process, which avoids damages caused by the temperature increases and enables the irradiation of cooled and frozen products.

The disadvantages of the use of irradiation are high initial cost and difficulty in establishing the right doses [102].

Although these are all the benefits, there are several barriers that still persist and prevent irradiated foods from reaching a wide commercialization, mainly related to the cost and consumer resistance due to lack of information [103].

4. Conclusions

The main concern regarding to food safety of eggs is related to the presence of pathogenic microorganism. As an attempt to reduce problems resulting from egg contamination, in addition to prolonging shelf life and ensuring greater safety for consumers, they are subjected to thermal and nonthermal processes. Pasteurization is a thermal method widely used, and it has efficiency in the decontamination of this food, but the use of heat can alter the nutritional quality, flavor, and texture of the products. The alternatives to the traditional pasteurization are the new technologies. In this way, alternative methods of food preservation that minimize the likelihood of outbreaks of food poisoning leading to improvements in food safety are of great importance. The new technologies are efficient in reducing the microbial load, if well used, cause minor alterations in the nutritional and organoleptic properties, contributing to the offer of a fresher product, besides being safe from the microbiological point of view. In addition, the combination of pasteurization methods with other alternative methods needs to be studied in order to provide quality to eggs and their products without affecting their properties and functionalities.

Acronyms and abbreviations

FDA	Food and Drug Administration
HHP	high hydrostatic pressure
HELP	high-intensity electric field
O_3	ozone
KI	potassium iodide
PEF	pulsed electric field
RF	radiofrequency
UV-C	short-wave ultraviolet radiation
UV	ultraviolet

Author details

Poliana Mendes de Souza[1*], Regiane de Melo[1], Miriam Aparecida de Aguilar Santos[2], Fabiana Regina Lima[1] and Kássia Héllen Vieira[1]

1 Universidade Federal dos Vales do Jequitinhonha e Mucuri, Diamantina, Brazil

2 Universidade Federal de Lavras, Lavras, Brazil

*Address all correspondence to: poliana.souza@ict.ufvjm.edu.br

References

[1] Sfaciotte RAP, Barbosa MJB, Wosiacki SR, Cardozo RM, Martins RR. Efeito do período de armazenamento, local e tipo de tratamento sobre a qualidade de ovos brancos para consumo humano. Pubvet. 2014;**8**:1-23

[2] FAO—Food and Agricultural Organization. Agribusiness Handbook. Poultry Meat & Eggs. Roma: FAO Investment Centre Division; 2015

[3] Potter NN, Hotchkiss JH. Food science. In: Potter NN, Hotchkiss JHF, editors. Food Science. 5th ed. Estados Unidos da América: Chapman & Hall; 1995. pp. 316-344

[4] Souza PM, Muller A, Beniaich A, Manyer-Miebach E, Oehlke K, Stahl M, et al. Functional properties and nutritional composition of liquid egg products treated in a coiled tube UV-C reactor. Innovative Food Science & Emerging Technologies. 2015;**32**:156-164. DOI: 10.1016/j.ifset.2015.09.004

[5] Mendes AA. Alimento perfeito. Avicultura Industrial. 2002;**3**:32-33

[6] Souza PM. Study of short-wave ultraviolet treatments (UV-C) as a non-thermal preservation process for liquid egg products [tese]. Valência: Universidade Politécnica de Valência; 2012

[7] USDA-AMS. USDA egg-grading manual. Agricultural Marketing Service, Agricultural handbook. 2000;**75**:1-56

[8] Foregeding EA, Luck PJ, Davis JP. Factors determining thephysical properties of protein foams. Food Hydrocolloids. 2006;**20**:284-292. DOI: 10.1016/j.foodhyd.2005.03.014

[9] Araújo WMC, Montebello NP, Botelho RBA, Borgo LA. Alquimia dos Alimentos. Brasília: Senac; 2011. 500 p

[10] Telis RJ, Thomas CEP, Bernardi M, Telis VRN, Gabas AL. Rheological properties and fluid dynamics of egg yolk. Journal of Food Engineering. 2006;**74**:191-197. DOI: 10.1016/j.jfoodeng.2005.01.044

[11] Pombo CR. Efeitos do tratamento térmico de ovos inteiros na perda de peso e características da qualidade interna [dissertação]. Niterói: Universidade Federal Fluminense; 2003

[12] Jones DR, Musgrove MT, Northcutt JK. Variations in external and internal microbial populations in shell eggs during extended storage. Journal of Food Protection. 2004;**67**:2657-2660. DOI: 10.4315/0362-028X-67.12.2657

[13] De Reu K, Grijspeerdt K, Messens W, Heyndrickx M, Uyttendaele M, Debevere J, et al. Eggshell factors influencing eggshell penetration and whole egg contamination by different bacteria, including Salmonella Enteritidis. International Journal of Food Microbiology. 2006;**112**:253-260. DOI: 10.1016/j.ijfoodmicro.2006.04.011

[14] Aragon-Alegro LC, Souza KLO, Sobrinho PSC, Landgraf M, Destro MT. Avaliação da qualidade microbiológica de ovo integral pasteurizado produzido com e sem a etapa de lavagem no processamento. Ciência e Tecnologia de Alimentos. 2005;**25**:618-622. DOI: 10.1590/S0101-20612005000300036

[15] Cunha FLC, Calixto FAA, Carneiro CS, Carrijo KF. Processamento, pasteurização, desidratação e outros processos similares de conservação de ovos de consumo. Pubvet. 2012;**31**:1-27

[16] Keerthirathne TP, Ross K, Fallowfield H, Whiley H. Reducing risk of salmonellosis through egg decontamination processes. International Journal of Environmental

Research and Public Health. 2017;**14**: 1-10. DOI: 10.3390/ijerph14030335

[17] Keener KM. Shell egg pasteurization. In: Hester PY, editor. Egg Innovations and Strategies for Improvements. 1st ed. Cambridge: Academic Press; 2017. pp. 165-175. DOI: 10.1016/B978-0-12-800879-9.00016-0

[18] Shenga E, Singh R, Yadav A. Effect of pasteurization of shell egg on its quality characteristics under ambient storage. Journal of Food Science and Technology. 2010;**47**:420-425. DOI: 10.1007/s13197-010-0069-2

[19] Patrignani F, Vannini L, Kamdem SLS, Hernando I, Molés RM, Guerzoni ME, et al. High pressure homogenization vs heat treatment: Safety and functional properties of liquid whole egg. Food Microbiology. 2013;**36**:63-69. DOI: 10.1016/j.fm.2013.04.004

[20] James C, Lechevalier V, Ketteringham L. Surface pasteurisation of shell eggs. Journal of Food Engineering. 2002;**53**:193-197. DOI: 10.1016/S0260-8774(01)00156-X

[21] Muriana PM, Hou H, Singh RK. A flow-injection system for studying heat inactivation of listeria monocytogenes and Salmonella Enteritidis in liquid whole egg. Journal of Food Protection. 1996;**59**:121-126. DOI: 10.4315/0362-028X-59.2.121

[22] Geveke DJ, Gurtler JB, Jones DR, Bigley AB. Inactivation of salmonella in shell eggs by hot water immersion and its effect on quality. Journal of Food Science. 2016;**8**:709-714. DOI: 10.1111/1750-3841.13233

[23] Liu YC, Chen TH, Wu YC, Lee YC, Tan FJ. Effects of egg washing and storage temperature on the quality of eggshell cuticle and eggs. Food Chemistry. 2016;**211**:687-693. DOI: 10.1016/j.foodchem.2016.05.056

[24] Pasquali F, Fabbri A, Cevoli C, Manfreda G, Franchini A. Hot air treatment for surface decontamination of table eggs. Food Control. 2010;**21**:431-435. DOI: 10.1016/j.foodcont.2009.07.003

[25] Zhang W, Liu F, Nindo C, Tang J. Physical properties of egg whites and whole eggs relevant to microwave pasteurization. Journal of Food Engineering. 2013;**1**:62-69. DOI: 10.1016/j.jfoodeng.2013.03.003

[26] Lakins G, Alvarado CZ, Thompson LD, Brashears MT, Brooks JC, Brashears MM. Reduction of *Salmonella enteritidis* in shell eggs using directional microwave technology. Poultry Science. 2008;5: 985-991. DOI: https: 10.3382/ps.2007-00393

[27] Salazar-González C, San Martín-González MF, López-Malo A, Sosa-Morales ME. Recent studies related to microwave processing of fluid foods. Food and Bioprocess Technology. 2012;**5**:31-46. DOI: 10.1007/s11947-011-0639-y

[28] Dev SR, Gariepy Y, Orsat V, Raghavan VG. FDTD modeling and simulation of microwave heating of in-shell eggs. Progress in Electromagnetics Research. 2010;**13**:229-243. DOI: 10.2528/PIERM10072609

[29] Dev SRS, Raghavan GSV, Gariepy Y. Dielectric properties of egg components and microwave heating for in-shell pasteurization of eggs. Journal of Food Engineering. 2008;**2**:207-2014. DOI: 10.1016/j.jfoodeng.2007.09.027

[30] Piyasena P, Dussault C, Koutchma T, Ramaswamy HS, Awuah GB. Radio frequency heating of foods: Principles, applications and related properties—A review. Critical Reviews in Food Science and Nutrition. 2003;**43**:587-606. DOI: 10.1080/10408690390251129

[31] Marra F, Zhang L, Lyng JG. Radio frequency treatment of foods: Review of recente advances. Journal of Food Engineering. 2009;**91**:497-508. DOI: 10.1016/j.jfoodeng.2008.10.015

[32] Kannan S, Dev SR, Gariépy Y, Raghavan VG. Effect of radiofrequency heating on the dielectric and physical properties of eggs. Progress in Electromagnetics Research B. 2013;**51**:201-220. DOI: 10.2528/PIERB13031812

[33] Lau SK, Thippareddi H, Subbiah J. Radiofrequency heating for enhancing microbial safety of shell eggs immersed in deionized water. Journal of Food Science. 2017;**82**:2933-2943. DOI: 10.1111 / 1750-3841.13965

[34] Geveke DJ, Bigley ABW, Brunkhorst CD. Pasteurization of shell eggs using radio frequency heating. Journal of Food Engineering. 2017;**193**:53-57. DOI: 10.1016/j.jfoodeng.2016.08.009

[35] Lau SK, Thippareddi H, Jones D, Negahban H, Subbiah J. Challenges in radiofrequency pasteurization of shell eggs: Coagulation rings. Journal of Food Science. 2016;**81**:2492-2502. DOI: 10.1111/1750-3841.13440

[36] Aquino JS, Silva JA, Prado JP, Cavalheiro JMO. Análise dos constituintes de gema de ovo de avestruz desidratada por meio de duas metodologias de secagem. Revista Instituto Adolfo Lutz. 2008;**67**:190-195

[37] Terrone HC, Jesus JM, Artuzo LT, Ventura LV, Santos RF, Damy-Benedetti PC. Liofilização. Revista Científica Unilago. 2013;**1**:271-284

[38] Grassi TLM, Ponsano EHG. Desidratação de gemas de ovos por secagem por atomização em diferentes temperaturas. Pesquisa Agropecuária Brasileira. 2015;**50**:1186-1191. DOI: 10.1590/S0100-204X2015001200008

[39] Wenzel M, Seuss-Baum I, Schlich E. Influence of pasteurization, spray- and freeze-drying, and storage on the carotenoid content in egg yolk. Journal of Agricultural and Food Chemistry. 2010;**58**:1726-1731. DOI: 10.1021/jf903488b

[40] Chen C, Chi YJ, Xu W. Comparisons on the functional properties and antioxidant activity of spray-dried and freeze-dried egg white protein hydrolysate. Food Bioprocess Technology. 2012;**5**:2342-2352. DOI: 10.1007/s11947-011-0606-7

[41] Eggs Safety Center. Pathogens [Internet]. 2010. Available from: http://www.eggsafety.org/consumers/pathogens [Accessed: Jun 20, 2018]

[42] Schmidt-Lorenz W. Collection of Methods for the Microbiological Examination of Foods. Weinheim: Verlag Chemie; 1983. pp. 15.1-15.22

[43] Cunningham FE. Egg-product pasteurization. In: Stadelman WJ, Cotterill OJ, editors. Egg Science and Technology. New York: Food Products Press; 1995. pp. 289-322

[44] Yuceer M, Caner C. Extending shelf life of egg using active packaging and novel preserving methods [Internet]. Available from: https://www.researchgate.net/project/Extending-Shelf-Life-of-Egg-Using-Active-Packaging-and-Novel-Preserving-Methods [Accessed: Jun 15, 2018]

[45] Marco-Molés R, Maria A, Rojas-Graü IH, P'erez-Munuera I, Soliva-Fortuny R, Mart'ın-Belloso O. Physical and structural changes in liquid whole egg treated with high-intensity pulsed electric fields. Journal of Food Science. 2011;**76**:257-264. DOI: 10.1111/j.1750-3841.2010.02016.x

[46] Souza PM, Müller A, Fernández A, Stahl M. Microbiological efficacy in

liquid egg products of a UV-C treatment in a coiled reactor. Innovative Food Science and Emerging Technologies. 2014;**21**:90-98. DOI: 10.1016/j.ifset.2013.10.017

[47] Koutchma T. Adapting High Hydrostatic Pressure (HPP) for Food Processing Operations. San Diego: Elsevier; 2014. 73 p. ISBN: 978-0-12-420091-3

[48] Balasubramaniam VM, Martínez-Monteagudo SI, Gupta R. Principles and application of high pressure-based technologies in the food industry. Annual Review of Food Science and Technology. 2015;**6**:435-462. DOI: 10.1146/annurev-food-022814-015539

[49] Health Canada. 2015. Novel food information—High pressure processing (HPP)-treated egg salad, egg dips, and egg spreads. Available from: http://www.hc-sc.gc.ca/fn-an/gmf-agm/appro/hpp_egg-oeuf_uhp-eng.php [Accessed: Sep 10, 2018]

[50] Naderi N, House JD, Pouliot Y, Doyen A. Effects of high hydrostatic pressure processing on hen egg compounds and egg products. Comprehensive Reviews in Food Science and Food Safety. 2017;**16**:707-720. DOI: 10.1111/1541-4337.12273

[51] NFL (National Food Lab). 2013. High pressure processing: Insights on technology and regulatory requirements. Available from: http://www.thenfl.com/wp-content/uploads/High-Pressure-Processing-Insights_20131.pdf [Accesed: Sep 10, 2018]

[52] Rendueles E, Omer MK, Alvseike O, Alonso-Calleja C, Capita R, Prieto M. Microbiological food safety assessment of high hydrostatic pressure processing: A review. LWT—Food Science and Technology. 2011;**44**:1251-1260. DOI: 10.1016/j.lwt.2010.11.001

[53] Monfort S, Ramos S, Meneses N, Knorr D, Raso J, Álvarez I. Design and evaluation of a high hydrostatic pressure combined process for pasteurization of liquid whole egg. Innovative Food Science and Emerging Technologies. 2012;**14**:1-10. DOI: 10.1016/j.ifset.2012.01.004

[54] Oxen P, Knorr D. Baroprotective effects of high solute concentrations against inactivation of Rhodotorula rubra. Lebensmittel Wissenschaft und Technologie. 1993;**26**:220-223

[55] Palou E, Lopez-Malo A, Barbosa-Canovas GV, Welti-Chanes J, Swanson BG. Effect of water activity on high hydrostatic pressure inhibition of Zygosaccharomyces bailii. Letters in Applied Microbiology. 1997;**24**:417-420

[56] Németh C, Dalmadi I, Mráz B, Friedrich L, Zeke I, Juhász R, et al. Effect of high pressure treatment on liquid whole egg. High Pressure Research. 2012;**32**:330-336. DOI: 10.1080/08957959.2012.687050

[57] Tóth A, Németh CS, Palotás P, Surányi J, Zeke I, Csehi B, et al. HHP treatment of liquid egg at 200-350 MPa. Journal of Physics: Conf. Series. 2017;**950**:042008. DOI: 10.1088/1742-6596/950/4/042008

[58] Cortés C, Frigola A, Esteve MJ. Color of orange juice treated by high intensity pulsed electric fields during refrigerated storage and comparison with pasteurized juice. Food Control. 2008;**19**:151-158. DOI: 10.1016/j.foodcont.2007.03.001

[59] Pettit B, Ritz M, Federighi M. Nouveaux Treatments Physiques de Conservation des Aliments: Revue Bibliographique. Revue de Médecine Vétérinaire. 2002;**53**:547-556

[60] Góngora-Nieto MM, Sepúlveda DR, Pedrow P, Barbosa-Cánovas GV, Swanson BG. Food processing by pulsed

electric fields: Treatment delivery, inactivation level, and regulatory aspects. Lebensmittel-Wissenschaft and Technologie. 2002;**35**:375-388. DOI: 10.1006/fstl.2001.0880

[61] Zhang Q, Quin BL, Barbosa-Cánovas GV, Swanson BG. Inactivation of *E. coli* for food pasteurization by high-strength pulsed electric fields. Jounal of Food Process Preservation. 1995;**19**:103-118. DOI: 10.1111/j.1745-4549.1995.tb00281.x

[62] Fani M. Novas técnicas de conservação. Conservação de Alimentos. Aditivos & Alimentos. 2010;**73**:40-49

[63] Palomeque LA, Gongora-Nieto MM, Bermudez AS, Barbosa-Canovas GV, Swanson BG. Nonthermal inactivation of endoproteases by pulsed electric fields technology in pulsed electric fields in food processing. In: Barbosa-Canovas GV, Zhang QH, editors. Pulsed Electric Fields in Food Processing. Lancaster, PA: Technomic Publishing Co. Inc.; 2001. pp. 135-148

[64] Yogesh K. Pulsed electric field processing of egg products: A review. Journal of Food Science and Technology. 2016;**53**:934-945. DOI: 10.1007/s13197-015-2061-3

[65] Gongora-Nieto MM, Seignour L, Riquet P, Davidson PM, Barbosa-Canovas GV, Swanson BG. Nonthermal inactivation of *Pseudomonas fluorescens* in liquid whole egg in pulsed electric fields in food processing. In: Barbosa-Canovas GV, Zhang QH, editors. Pulsed Electric Fields in Food Processing. Lancaster, PA: Technomic Publishing Co. Inc.; 2001. pp. 193-211

[66] Zhao W, Yang R, Tang Y, Lu R. Combined effects of heat and PEF on microbial inactivation and quality of liquid egg whites. International Journal of Food Engineering. 2007;**3**:1-20. DOI: 10.2202/1556-3758.1256

[67] Mine Y. Recent advances in the understanding of egg white protein functionality. Trends in Food Science & Technology. 1995;**6**:225-231. DOI: 10.1016/S0924-2244(00)89083-4

[68] Azerêdo G, Oliveira FLN, Faro ZP. Pulsos Elétricos na preservação de Alimentos: Fatores Críticos Na inativação Microbiana e Efeitos Sobre os Constituintes Alimentares. Boletim do Centro de Pesquisa de Processamento de Alimentos. 2008;**26**:171-178

[69] Lado BH, Yousef AE. Alternative food-preservation technologies: Efficacy and mechanisms. Microbes and Infection. 2002;**4**:433-440

[70] Knorr D, Geulen M, Grahl T, Sitzmann W. Food application of high electric field pulses. Trends in Food Science and Technology. 1994;**5**:71-75. DOI: 10.1016/0924-2244(94)90240-2

[71] Jansen A. Pioneering with pulsed electric field. Holland Food Innovations. 2015;**1**:6-9

[72] Khadre MA, Yousef AE, Kim JG. Microbiological aspects of ozone applications in food: A review. Journal of Food Science. 2001;**66**(9):1242-1252

[73] Poppendieck D, Hubbarda H, Warda M, et al. Ozone reactions with indoor materials during building disinfection. Atmospheric Environment. 2007;**41**:3166-3176

[74] Horváth M, Bilitzky L, Hüttner J. Cap. 3: Bactericidal, sterilizing and other effects in lower organisms. In: Ozone. Budapest: Science; 1985. pp. 69-74

[75] Braun PG, Fernandez N, Fuhrmann H. Investigations on the effect of ozone as a disinfectant of egg surfaces. Ozone: Science & Engineering: The Journal of the International Ozone Association. 2011;**33**:374-378. http://dx.doi.org/10.1080/01919512.2011.589359

[76] Kasper MA, Guisso CA, Bonavigo A, Schropfer DL, Bassani MT, Ebling PD. Aplicação de Ozônio em Ovos para Consumo [Internet]. Available from: https://eventos.uceff.edu.br/eventosfai_dados/artigos/agrotec2017/644.pdf [Accessed: Jul 6, 2018]

[77] Yüceer M, Adayb SM, Canerb C. Ozone treatment of shell eggs to preserve functional quality and enhance shelf life during storage. Science of Food and Agriculture. 2016;**96**:2755-2763. DOI: 10.1002/jsfa.7440

[78] Mahapatra AK, Muthukumarappan K, Julson JL. Applications of ozone, bacteriocins and irradiation in food processing: A review. Critical Reviews in Food Science and Nutrition. 2005;**45**(6):447-461

[79] Kim JG, Yousef AE, Dave S. Application of ozone for enhancing the microbiological safety and quality of foods: A review. Journal of Food Protection. 1999;**62**:1071-1087

[80] Alexandre FA, Faria JAF, Cardoso CF. Avaliação da eficiência da radiação ultravioleta na esterilização de embalagens plásticas. Ciência e Agrotecnologia. 2008;**32**(5):1524-1530

[81] Krishnamurthy K, Demirci A, Irudayaraj JM. Staphylococcus aureus inactivation using pulsed UV light for continuous milk treatment. In: ASAE Annual Meeting; Florida. 2005. pp. 1-11. DOI: 10.13031/2013.19590

[82] Gottselig SM. Microbial reduction on eggshell surfaces by the use of hydrogen peroxide and ultraviolet light [Master of science, thesis]. Texas A&M University. 2011. 91 p

[83] Guedes AMM, Novello D, Mendes GMP, Cristianini M. Tecnologia de ultravioleta para preservação de alimentos. Boletim do Centro de Pesquisa de Processamento de Alimentos. 2009;**27**(1):59-70

[84] Shama G. Process challenges in applying low doses of ultraviolet light to fresh produce for eliciting beneficial hormetic responses. Postharvest Biology and Technology. 2007;**44**(1):1-8

[85] Geveke DJ. UV inactivation of *E. coli* in liquid egg white. Food Bioprocess Technology. 2008;**1**:201-206. DOI 10.1007/s11947-008-0070-1

[86] Souza PM, Fernandez A. Consumer acceptance of UV-C treated liquid egg products and preparations with UV-C treated eggs. Innovative Food Science and Emerging Technologies.. 2012;**14**:107-114. DOI: 10.1016/j.ifset.2011.12.005

[87] Silva GRD, Menezes LDM, Lanza IP, Oliveira DD, Silva CA, Klein RWT, et al. Evaluation of the alpha-amylase activity as an indicator of pasteurization efficiency and microbiological quality of liquid whole eggs. Poultry Science. 2017;**96**: 3375-3381. DOI: 10.3382/ps/pex108

[88] Lagunas-Solar MC, Pina C, Macdonald JD, Bolkan L. Development of pulsed UV light processes for surface fungal disinfection of fresh fruits. Journal of Food Protection. 2006;**69**:376-384

[89] Sevcan U, Mehmte RA, Handan AB, Canam T. Use of UV-C radiation as a non-thermal process for liquid egg products (LEP). Journal of Food Engineering 2008;**85**:561-568. DOI: 10.1016/j.jfoodeng.2007.08.017

[90] Cheis D. Desinfecção de água e efluentes com raios ultravioleta. Revista TAE [Internet]. 2013. Available form: http://www.revistatae.com.br/6102-noticias [Accessed: Jul 8, 2018]

[91] Chang JCH, Ossoff SF, Lobe DC, Dorfman MH, Dumais CM, Qualls RG, et al. UV inactivation of pathogenic and indicator microorganisms. Apllied and Environmental Microbiology. 1985;**49**:1361-1365

[92] Morgan R. UV 'green' light disinfection. Dairy Industries International. 1989;**54**:33-35

[93] Barancelli GV, Marin JGP, Porto E. Salmonella em ovos: Relação entre produção e consumo seguro. Segurança Alimentar e Nutricional, Campinas. 2012;**19**:73-82

[94] Marquis REE, Baldeck JD. Sporicidal interactions of ultraviolet irradiation and hydrogen peroxide related to aseptic technology. Chemical Engineering and Processing. 2007;**46**:547-553. DOI: 10.1016/j.cep.2006.07.009

[95] Gennadios A, Handa A, Rhim J, Handa M. Ultraviolet radiations affects physical and molecular properties of soy protein films. Journal of Food Science. 1998;**63**:1-4. DOI: 10.1111/j.1365-2621.1998.tb15714.x

[96] Fellows PJ. Tecnologia do Processamento de Alimentos—Princípios e práticas. Porto Alegre: Artmed; 2006. 602 p. ISBN: 978-85-363-0652-0

[97] Lima LS. Radiação gama. Revista de Ciência Elementar. 2014;**2**:1-2

[98] Centro de Energia Nuclear na Agricultura (CENA). Divulgação da Tecnologia da Irradiação de Alimentos e Outros Materiais—Equipamentos para Irradiação de Alimentos e/ou Outros Materiais [Internet]. 2002. Available from: http://www.cena.usp.br [Accessed: Sep 1, 2018]

[99] Caldeira H, Bello A, Gomes J. Aplicações de radiações ionizantes [Internet]. 2010. Available from: http://fisicaradioactiva.blogspot.com/p/aplicacoes-da-radiacao.html= [Accessed: Aug 7, 2017]

[100] Franco BDGM, Landgraf M. Microbiologia dos Alimentos. São Paulo: Ed. Atheneu; 2008. 182 p

[101] Harder MNC, Arthur V. Effects of gamma radiation for microbiological control in eggs. In: Hester PY, editor. Egg Innovations and Strategies for Improvements. 1st ed. Cambridge: Academic Press; 2017. pp. 165-175. DOI: 10.1016/ B978-0-12-800879-9.00016-0

[102] Hallman GJ. Ionizing gradiation quarantine treatments against tephritic fruit flies. Postharvest Biology and Technology. 1999;**16**:93-106. DOI: 10.1016/S0925-5214(99)00012-5

[103] Ornellas CBD, Gonçalves MPJ, Silva PR, Martins RT. Atitudes do consumidor frente à irradiação de alimentos. Ciência e Tecnologia de Alimentos. 2006;**26**:211-213. DOI: 10.1590/S0101-20612006000100033

Section 3

Food Fortification

Chapter 6

Food Fortification through Innovative Technologies

Vlaic Romina Alina, Mureşan Crina Carmen, Muste Sevastita, Mureşan Andruţa, Muresan Vlad, Suharoschi Ramona, Petruţ Georgiana and Mihai Mihaela

Abstract

The chapter aims to approach food fortification naturally as a result of the need for nutritional improvement and therefore underlines sustainable activities that would facilitate effective fortification. The need to fortify food is due to the close link between human, health, and food. The WHO and FAO and other internationally recognized organizations have recognized that there are over 2 billion people worldwide suffering from a variety of micronutrient deficiencies. The interest in the fortification of foods is largely due to bioactive compounds, such as vitamins, minerals, sugars, organic acids, dietary fiber, phenolic compounds, essential amino acids, and antioxidants. The most effective and accessible way of securing the population with vitamins and micronutrients is to fortify additional food and consumer products daily. At the same time, the technology for the fortification of bakery products will also be presented.

Keywords: food, fortification, nutrients, advanced technologies

1. Introduction

In accordance with the *General Principles for the Addition of Essential Nutrients to Foods* [1] Codex, the term *fortification* or *enrichment* is synonymous to the addition of one or several essential nutritional elements for a food product, regardless of whether it is or not habitually contained in foods, toward the prevention or correction of proven deficiencies in one or more nutrients, either for the entire population or certain specific groups. The General Principles Codex further states as the primary condition mentioned in the fulfillment of any such fortification program to be the initial demonstration of the requirement for an increase in the nutrient intake essential to the population or to the target group. This demonstration can rely on real clinical or subclinical studies of deficiency, estimates of the low levels of nutrient intake, or possible deficiencies triggered by changes in standard food products [1].

Food fortification has been implemented for a long period of time in industrialized countries to achieve the successful control of vitamin A and D deficiencies, several B vitamins (thiamine, riboflavin, and niacin), as well as iodine and iron. Salt irrigation was introduced in the early 1920s both in Switzerland [2] and the United States [3] and has progressively expanded worldwide since salt iodine is presently used in most countries. Since the early 1940s, the fortification of cereal products with thiamine, riboflavin, and niacin has become a common practice. Margarine was fortified with vitamin A in Denmark, while milk was fortified with vitamin D in the United States. Foods for young children were fortified with iron, thus substantially reducing the risk of iron-deficiency anemia in this age group. In recent years, folic acid fortification has become widespread in America, a strategy undertaken by Canada and around 20 Latin American countries [4]. These approaches have proven to be effective in reducing the prevalence of many diseases due to deficiencies, such as goiter (iodine), xerophthalmia (vitamin A), rickets (vitamin D), and anemia (iron). Foods from around the world have begun to be fortified with calcium, iron, phosphorus, and vitamins (especially A, B, C, and D), depending on the chemical composition of the basic foods [5].

The need for food fortification is also presented by the WHO and FAO, among many other nationally recognized organizations. They have acknowledged the micronutrient deficiencies of more than 2 billion people, largely caused by a dietary deficiency of vitamins and minerals. The importance of these deficiencies for public health depends on their magnitude and spread, but the most vulnerable are pregnant women and young children, as fetal and child development, cognitive development, and resistance to infections are affected. Although the entire world population suffers from various nutritional deficiencies, people with low incomes are the most affected, particularly in developing countries, due to unsafe food consumption. Poverty, the lack of access to a variety of foods, and the lack of knowledge on appropriate dietary practices represent major drawbacks for socio-economic development while also contributing to a vicious circle of underdevelopment. As such, long-term effects on health, learning, and productivity are significant, while they also generate a high level of social and public costs from reduced work capacity due to high rates of illness and disability [4].

It was also in less industrialized countries that fortification has become more and more appealing in recent years, so much that the planned programs have advanced more rapidly throughout the implementation phase than initially estimated. Bearing in mind the success of the relatively long program to fortify sugar with vitamin A in Central America, where the prevalence of vitamin A deficiency has been greatly reduced, there are similar initiatives in other regions of the world. Currently, the first attempt to fortify sugar in South Africa is taking place in Zambia, and if successful, it will be implemented elsewhere. Darnton-Hill and Nalubola [5] identified at least 27 developing countries that could benefit from programs to fortify one or more foods.

According to research conducted by Svetlana [6], the most efficient and accessible way of providing the population with vitamins and micronutrients is their additional fortification using these substances in consumer food products and daily foods, especially flour and bakery products. The fortification of foods should not diminish their nutritional qualities and quality, especially not to substantially alter the taste or assimilation of other nutrients contained therein, not to reduce shelf life or to change product harmlessness characteristics.

2. Biofortification

People require at least 22 mineral elements for their well-being [7]. These micronutrients can be delivered through a proper diet. However, over 60% of the world's population is estimated to be iron-deficient(Fe), more than 30% are zinc-deficient (Zn), while 30% are iodine-deficient and 15% selenium-deficient (Se). In addition, calcium (Ca), magnesium (Mg), and copper (Cu) deficiencies are common in many developed and developing countries [8]. These deficits influence many biochemical pathways directly or indirectly [9]. This is due to the production of crops in areas with low mineral bioavailability and/or (sporadic) crop consumption with inherent tissue mineral concentrations, accompanied by the lack of fish or food products [7] or low-micronutrient crops [10]. At present, mineral malnutrition is considered to be one of the most serious global challenges that is active but avoidable [11]. Mineral malnutrition can be addressed by dietary diversification, mineral supplements, food fortification and/or increased mineral concentration in edible crops (biofortification).

The term *biofortification* refers to the increase of the micronutrient concentration in the edible part of the plant and can be achieved both by using fertilizers and by stimulating the absorption of these minerals in the plant [12], while the use of micronutrient fertilizers was effective in increasing the micronutrient content in plants [13].

Practically, the fortification of food is closely linked with the biofortification of plants (raw materials for food); the higher the nutritional value of the raw material, the smaller the need for fortification. Thus, from increasing the productivity of the cultivated varieties nowadays, the necessity of obtaining varieties with high nutritional value has become a very important task for agronomists, as almost half of the world population suffers from deficiencies in zinc, iron, and selenium [14].

In some countries, selenium wheat biofortification is achieved by using selenium-based fertilizers, as wheat is considered a major source of selenium for our daily diet [15]. As such, selenium is considered to be an essential micronutrient for the human body, which can reduce the risk of degenerative diseases, including cancer [16]. A study by Wu [17] based on the consumption of common wheat biscuits biofortified with selenium through fertilization showed that although the levels of selenium in blood plasma increased, there were no significant changes of degenerative disease-specific biomarkers and of the health condition in general. Another method of biofortification that aimed at the development of wheat varieties enriched with some micronutrients was based on interactions between genotypes and environmental factors and led to good iron and zinc levels in humans [18, 19].

3. Food fortification

Government agencies, as well as food policy makers, support food enrichment and fortification to different degrees, in order to lower the deficiency rate in populations on a large scale, this proving to be an efficient approach [20]. It is estimated that 2 billion people worldwide suffer from micronutrient malnutrition [21]. From a public health standpoint, it is estimated that micronutrient deficiencies account for approximately 7.3% of global diseases.

The most important elements used for food fortification are:

3.1 Iron

The greatest part of the iron in the human body is found in erythrocytes as hemoglobin, where its main function is to carry oxygen from the lungs onto the tissues. Iron deficiency causes anemia, the most common and widespread nutritional disorder in the world and a public health problem in both industrialized and nonindustrialized countries [22]. As a component of myoglobin, a protein that supplies oxygen to the muscles, iron supports metabolism [23]. Additionally, iron is necessary for growth, development, normal cellular function, and the synthesis of hormones and the connective tissue [23]. Dietetic iron has two main forms: heme and nonheme [24]. Iron-fortified plants and foods contain only nonheme iron, while meat, seafood, and poultry contain both heme and nonheme iron [23].

People normally lose small amounts of iron through the urine, feces, gastrointestinal tract, and skin. Losses are higher in menstrual women due to blood loss. Hepcidin, a circulating peptide hormone, is the key element in regulating iron absorption and iron distribution in the body, including plasma [23–25].

Hemoglobin concentrations lower than 13 g/dL in men and 12 g/dL in women indicate the incidence of iron-deficiency anemia (IDA) [26].

The Recommended Dietary Allowance (RDA) is shown in **Table 1**; RDA for vegetarians is 1.8 times higher than for those who eat meat. This is because iron from meat is more bioavailable than the iron in herbal foods, while meat, poultry, and seafood increase iron absorption through nonheme [26].

The richest sources of heme iron in the diet include lean meats and seafood. Dietary nonheme iron sources include nuts, beans, vegetables, and fortified cereal products. In the United States, about half of the food is derived from bread, cereals, and other cereal products [23, 26]. Breast milk contains highly bioavailable iron but in quantities that are not sufficient to meet the needs of infants over 4–6 months [23, 32]. In the United States, Canada, and many other countries, wheat and other types of flour are fortified with iron [33]. Also, infant formulas are fortified with 12 mg iron per liter [32].

According to EFSA average iron intake ranged between 2.6 and 6.0 mg/day (0.9–1.9 mg/MJ) in infants (< 1 year, four surveys), between 5.0 and 7.0 mg/day (1.2–1.6 mg/MJ) in children aged 1 to <3 years (five surveys), between 7.5 and 11.5 mg/day (1.1–1.7 mg/MJ) in children aged 3 to <10 years (seven surveys), between 9.2 and 14.7 mg/day (1.1–1.7 mg/MJ) in children aged 10 to <18 years (seven surveys), and between 9.4 and 17.9 mg/day (1.2–2.1 mg/MJ) in adults (≥ 18 years) (eight surveys). Average daily intakes were in most cases slightly higher in males (Appendix D) than in females (Appendix E), mainly owing to larger quantities of food consumed per day [34]. Tolerable upper intake levels for iron fall between 40 and 45 mg [26].

A study conducted in the United States, for example, revealed that iron supplements during pregnancy have reduced the number of premature births or the incidence of low birth weight [35]. In Vietnam, fortifying fish sauce with iron and consuming 10 ml per day of the sauce fortified with 100 mg iron (as NaFeEDTA) for 100 ml significantly improve the iron deficiency of the group of women tested, compared to the placebo group after just 6 months [36]. In China, a series of studies have been carried out to evaluate the efficacy of fortifying soy sauce with iron (as NaFeEDTA). The daily consumption of 5 mg or 20 mg of iron in fortified sauce has been reported to be very effective in the treatment of iron-deficiency anemia in children; positive effects were recorded within 3 months of starting the study or 6 months on a sample of 10,000 children and women suffering from anemia [37]. In a South African iron deficiency group, fortifying the curry powder with NaFeEDTA led to significant improvements in blood hemoglobin and the prevalence of iron-deficiency anemia in women declined from 22% to just 5% during the 2 years of study [38].

3.2 Vitamin A

Vitamin A is the name of a group of retinoids soluble in fats, including retinol, retinal, and retinyl esters. Vitamin A represents an essential nutrient, thus necessary in small amounts for the normal functioning of the visual system in

Age/nutrient		Iron (mg/d)	Vitamin A (μg/d)	Iodine (μg/d)	Folate (μg/d)	Vitamin B12 (μg/d)	Thiamin (mg/d)	Riboflavin (mg/d)	Niacin (mg/d)	Vitamin B6 (mg/d)	Vitamin C (mg/d)	Vitamin D (μg/d)	Calcium (mg/d)	Selenium (μg/d)	Total fiber (g/d)	Protein (g/d)	Fat (g/d)
Infants	0–6 months	0.27	400	110	65	0.4	0.2	0.3	2	0.1	40	10	200	15	ND	9.1	31
	6–12 months	11	500	130	80	0.5	0.3	0.4	4	0.3	50	10	260	20	ND	11	30
Children	1–3 years	7	300	90	150	0.9	0.5	0.5	6	0.5	15	15	700	20	19	13	ND
	4–8 years	10	400	90	200	1.2	0.6	0.6	8	0.6	25	15	1000	30	25	19	ND
Males	9–13 years	8	600	120	300	1.8	0.9	0.9	12	1	45	15	1300	40	31	34	ND
	14–18 years	11	900	150	400	2.4	1.2	1.3	16	1.3	75	15	1300	55	38	52	ND
	19–30 years	8	900	150	400	2.4	1.2	1.3	16	1.3	90	15	1000	55	38	56	ND
	31–50 years	8	900	150	400	2.4	1.2	1.3	16	1.3	90	15	1000	55	38	56	ND
	51–70 years	8	900	150	400	2.4	1.2	1.3	16	1.7	90	15	1000	55	30	56	ND
	70+ years	8	900	150	400	2.4	1.2	1.3	16	1.7	90	20	1200	55	30	56	ND
Females	9–13 years	8	600	120	300	1.8	0.9	0.9	12	1	45	15	1300	40	26	34	ND
	14–18 years	15	700	150	400	2.4	1	1	14	1.2	65	15	1300	55	26	46	ND
	19–30 years	18	700	150	400	2.4	1.1	1.1	14	1.3	75	15	1000	55	25	46	ND
	31–50 years	18	700	150	400	2.4	1.1	1.1	14	1.3	75	15	1000	55	25	46	ND
	51–70 years	8	700	150	400	2.4	1.1	1.1	14	1.5	75	15	1200	55	21	46	ND
	70+ years	8	700	150	400	2.4	1.1	1.1	14	1.5	75	20	1200	55	21	46	ND
Pregnancy	14–18 years	27	750	220	600	2.6	1.4	1.4	18	1.9	80	15	1300	60	28	71	ND
	19–30 years	27	770	220	600	2.6	1.4	1.4	18	1.9	85	15	1000	60	28	71	ND
	31–50 years	27	770	220	600	2.6	1.4	1.4	18	1.9	85	15	1000	60	28	71	ND

Age/nutrient		Iron (mg/d)	Vitamin A (μg/d)	Iodine (μg/d)	Folate (μg/d)	Vitamin B12 (μg/d)	Thiamin (mg/d)	Riboflavin (mg/d)	Niacin (mg/d)	Vitamin B6 (mg/d)	Vitamin C (mg/d)	Vitamin D (μg/d)	Calcium (mg/d)	Selenium (μg/d)	Total fiber (g/d)	Protein (g/d)	Fat (g/d)
Lactation	14–18 years	10	1200	290	500	2.8	1.4	1.6	17	2	115	15	1300	70	29	71	ND
	19–30 years	9	1300	290	500	2.8	1.4	1.6	17	2	120	15	1000	70	29	71	ND
	31–50 years	9	1300	290	500	2.8	1.4	1.6	17	2	120	15	1000	70	29	71	ND
References		[26]	[26]	[26]	[27]	[27]	[27]	[27]	[27]	[27]	[28]	[29]	[29]	[30]	[31]	[31]	[31]

ND—not determined.

Table 1.
Nutrient RDA by age.

human beings, maintenance of cell growth function, epithelial cell integrity, immune function, and reproduction. Dietary vitamin A requirements are normally provided by a combination of preformed vitamin A (retinol), which is present in food of animal origin, and provitamin A carotenoids, which are derived from foods of plant origin, which have to be converted into retinol by tissues such as the intestinal mucosa and the liver, in order to be used by cells [39–41]. Vitamin A deficiency is based on a frequent evaluation of serum or plasma retinol [42], and a deficit affects visual function, as vitamin A status indicators have traditionally been based on visual changes, especially night blindness and xerophthalmia [43]. The World Health Organization (WHO) estimated that over 254 million preschool children worldwide have low serum retinol levels, and therefore these indicators may be considered clinically or subclinically deficiencies in vitamin A [42].

Two forms of vitamin A are available in the human diet: preformed vitamin A (retinol and its esterified form, retinyl ester) and provitamin A carotenoids [26, 39–41]. Preformed vitamin A is found in food from animal sources, including dairy products, fish, and meat (especially the liver). The most important provitamin A carotenoid is beta-carotene; other provitamin A carotenoids are alpha-carotene and beta-cryptoxanthin. The body transforms these pigments into vitamin A. Both provitamin A and preformed vitamin A should be metabolized intracellularly within the retina and retinoic acid, the active forms of vitamin A, to support the vital biological functions of the vitamin [40, 41]. Other carotenoids found in foods, such as lycopene, lutein, and zeaxanthin, are not turned into vitamin A.

Commonly, vitamin A deficiency occurs with a diet low in vitamin A sources (e.g., dairy products, eggs, fruits, and vegetables), poor nutritional status, and a high rate of infection, especially measles and diarrheal diseases. The best sources of vitamin A are foods of animal origin, especially liver, eggs, and dairy products containing vitamin A in the form of retinol, a form that can be easily used by the body. In fact, it is difficult for children to meet their vitamin A requirements if their diet is low in foods of animal origin [26, 40] and, similarly, if their diet is low in fat. Fruits and vegetables contain vitamin A in the form of carotenoids, the most important being β-carotene. In a mixed diet, the conversion rate of β-carotene to retinol is about 12: 1. The conversion of other carotenoids to retinol is less effective, the corresponding conversion rate being 24: 1. Different food preparation techniques, such as cooking, grinding, and adding oil, can improve the absorption of synthetic β-carotene food carotenoid in the oil, also widely used in vitamin A supplements. The latter has a conversion ratio of 2: 1 to retinol, while the synthetic β-carotene forms, commonly used to fortify foods, have a conversion ratio of 6: 1.

Analyses have shown that high-dose vitamin A supplements can reduce measles mortality by up to 50%. Another analysis found that improving vitamin A deficiencies, either by supplements or by fortification, decreased mortality by 23% in children aged 6 months to 5 years, irrespective of the cause [44]. Another study on a group of preschoolers who consumed 27 g of vitamin A-fortified margarine for 6 months reported a reduction in the prevalence of low serum retinol concentrations from 26 to 10% [45].

A plasma retinol concentration lower than 0.70 micromoles/L (or 20 micrograms/mcg/dL) reflects vitamin A deficiency in a population, whereas concentrations of 0.70–1.05 micromoles/L could be marginal in some people [5]. RDA for men and women is 900 and 700 μg retinol activity equivalent/day. Tolerable upper intake levels for preformed vitamin A in adults are set at 3000 μg [26].

3.3 Iodine

Iodine is present in the body in small amounts, mainly in the thyroid gland, while its only confirmed role is in the synthesis of thyroid hormones. Iodine deficiency is a major public health problem for populations around the world but especially for young children and pregnant women. In some areas, it is a significant threat to national, social, and economic development, as the devastating result of iodine deficiency is mental retardation: currently one of the main causes of cognitive impairment that can be prevented. This is the primary reason behind the current global mission to eliminate iodine deficiency-related disorders [4].

Iodine is an essential component of thyroid hormones, thyroxine, and triiodothyronine. Thyroid hormones regulate many important biochemical reactions, including protein synthesis and enzymatic activity, as critical determinants of metabolic activity [26, 46]. These are also necessary for an adequate development of the central nervous system and skeletal system in the fetus and infants [46].

Food iodine and iodine salt are present in several chemical forms, including sodium and potassium salts, inorganic iodine, iodate, and iodide, the reduced iodine form [47]. Iodine rarely appears as an element but rather as a salt. For this reason, it is called iodide and not iodine. Iodine is rapidly and almost completely absorbed in the stomach and duodenum. Iodate is reduced in the gastrointestinal tract and absorbed as iodide [26, 48]. When the iodide enters the circulation, the thyroid gland concentrates it in quantities adequate for thyroid hormone synthesis, and most of the remaining amount is excreted in the urine [26]. Healthy adults with sufficient iodine amounts exhibit approximately 15–20 mg of iodine, 70–80% of which are contained in the thyroid [49].

Average urinary iodine concentrations of 100–199 mcg/L in children and adults, 150–249 mcg/L in pregnant women, and > 100 mcg/L in nursing women indicate adequate iodine intake. Values lower than 100 mcg/L in children and adults who are not pregnant indicate insufficient iodine intake, although iodine deficiency is not classified as severe until the urinary iodine level is lower than 20 mcg/L [47].

One of the best sources of iodine is marine algae, but its content is variable according to species [48]. Other good sources include seafood, dairy products, cereal products, and eggs [50]. Iodine is also present in human breast milk [26, 48] and in infant formulas.

3.4 Folate (vitamin B9)

Folate (vitamin B9) is important in the synthesis and methylation of nucleotides that intervene in cell multiplication and tissue growth. Its role in protein synthesis and metabolism is closely interrelated to that of vitamin B12.

The main sources of dietary folate are leafy vegetables, fruits, yeast, and liver [4]. The total body content of folate is estimated at 10–30 mg; about half of this amount is stored in the liver [51, 52], while the rest in the blood and body tissues. A serum folate concentration is commonly used to assess folic acid status, with a value greater than 3 ng/ml being adequate [26, 51]. The follicular erythrocyte concentration above 140 ng/ml indicates an adequate folate status [26, 53], although some researchers have suggested that superior values are optimal for preventing neural tube defects [54].

Folate is naturally found in a wide variety of foods, including vegetables (especially leafy green vegetables), fruits and fruit juices, nuts, beans, peas, dairy products, poultry and meat, eggs, sea food, and cereal. Spinach, liver, yeast, asparagus, and brussels sprouts are among the foods with the highest levels of folic acid [52, 55].

In January 1998, the US Food and Drug Administration (FDA) began to require producers to add folic acid to enriched bread, cereals, flour, corn, pasta, rice, and other cereal products [56]. Since cereals are widely consumed in the United States, they have become very important contributors to folic acid intake for the American diet. The fortification program was designed to increase the folic acid intake in the United States by about 100 mcg/day but eventually succeeded an increase of about 190 mcg/day [57]. In April 2016, the FDA approved the voluntary addition of folic acid to corn meal at levels compatible with other enriched cereal products [58]. The Canadian Government has also called for the addition of folic acid to many cereals, including white flour, enriched pasta, and corn flour, as of November 1, 1998 [59]. Other countries, including Costa Rica, Chile, and South Africa, have also established compulsory programs for the folic acid fortification [60].

3.5 Vitamin B12

Vitamin B12 (cobalamin) is a cobalt-containing vitamin that is synthesized by microorganisms and exists in various chemical forms in food, particularly that of animal origin, such as milk, cheese and eggs, as well as artificially fortified foods [61]. B12 deficiency can cause neurological damage, megaloblastic anemia, increase of homocysteine in plasma, and possibly impairment of the immune function. In infants and young children, it can cause serious developmental delays [4]. Approximately 56% of an oral B12 dose of 1 mcg is absorbed, but absorption decreases drastically when the intrinsic factor capacity is exceeded (for 1–2 μm of vitamin B12) [62].

Vitamin B12 is systematically evaluated through B12 serum or plasma levels. Approximate values below 170–250 pg/ml (120–180 picomoles/L) for adults [26] indicate a vitamin B12 deficiency. Increased levels of methylmalonic acid (values>0.4 micromole/l) could be a more reliable indicator of vitamin B12 status, as they indicate a metabolic change that is highly specific to vitamin B12 deficiency [26, 32–64].

Vitamin B12 is naturally found in animal products, including fish, meat, poultry, eggs, milk, and dairy products. Generally, vitamin B12 is not present in plant foods, but fortified breakfast cereals are an available vitamin B12 source with high bio-availability for vegetarians. Some nutritional products from yeast also contain vitamin B12 [26, 65].

3.6 Other B vitamins (thiamine, riboflavin, niacin, and vitamin B6)

In recent years, due to increased evidence that some B vitamins can prevent the occurrence of developmental disorders, as well as chronic degenerative and neoplastic diseases, special attention has been paid to their possibilities of employment [66]. Moreover, vitamin B complexes are of utmost importance for energy metabolism. Specifically, thiamine (vitamin B1), riboflavin (vitamin B2), niacin (vitamin B3), and vitamin B6 are required for decarboxylation, transamination, acylation, oxidation, and reduction of substrates that are ultimately employed in energy consumption. One or several of these vitamins are also important for amino acid, fatty acids, cholesterol, steroid, and glucose synthesis [67]. In fact, thiamine plays a critical role in energy metabolism and hence in cell growth, development, and function. Important sources can be found in wheat germs and yeast extracts, the edible organs of most animals, legumes, and green vegetables [68]. The main sources of riboflavin are milk and dairy products, bread and bakery products, mixed meat-based foods, ready-to-eat cereals, and mixed grain-based foods [27, 69]; good sources of niacin are liver, meat and meat products, fish, peanuts, and whole grains [70]. The richest sources of vitamin B6 include fish, beef liver and other organs, potatoes, and other starchy vegetables, and fruits (other than citrus fruits) [27, 71]. Fortification of foods for nutrient recovery such as the vitamin B complex is effective in basic foods or spices [72], in wheat flour [73, 74], fish sauce [75], and rice [76]. RDAs are presented in **Table 1**.

3.7 Vitamin C

Vitamin C is a redox system composed of ascorbic acid and dehydroascorbic acid, which acts as an electron donor. Its main metabolic function is to maintain collagen formation. Additionally, it functions as an important antioxidant. Vitamin C is widely available in food of plant and animal origin, but the best sources are fresh fruits and vegetables and the edible organs of animals. However, since vitamin C is unstable when exposed to an alkaline environment or to oxygen, light, and heat, losses can be substantial during storage and cooking [4]. Foods fortified with vitamin C include milk and baby food [77], juices [78], jelly, and candies [79].

Acute vitamin C deficiency leads to scurvy. Scurvy evolution time varies with vitamin C levels, but signs may occur within 1 month after the decrease or absence of vitamin C consumption (below 10 mg/day) [80].

3.8 Vitamin D

Vitamin D deficiency can lead to musculoskeletal diseases such as rickets and osteomalacia, but vitamin D supplements and fortified foods can prevent extraskeletal disorders such as respiratory tract infections, asthma exacerbations, pregnancy complications, and premature death. Vitamin D has a unique metabolism as it is mainly synthesized in the skin under the influence of sunlight (i.e., ultraviolet radiation-B), while nutritional intake traditionally plays a relatively minor role. The recommended target concentrations range from ≥25 to ≥50 nmol/L (≥ 10–20 ng/ml), corresponding to a daily vitamin D dose of 10–20 μg (400–800 international units). Worldwide, vitamin D food fortification has already been introduced in the United States, Canada, India, and Finland with effective results [81]. As such, foods recommended to be fortified with vitamin D include milk, dairy products, and margarine [82].

Very few natural foods contain vitamin D. Fatty fish (such as salmon, tuna, and mackerel) and fish liver oils are among the best sources [28, 65]. Small amounts of vitamin D are found in beef liver, cheese and egg yolks, and certain mushroom varieties [83].

In the American diet, fortified foods provide most of the vitamin D intake [28, 83]. For example, almost all US milk is voluntarily fortified with 100 IU/cup. (In Canada, milk is fortified by law with 35–40 IU/100 ml, similar to margarine with ≥530 IU/100 g.) In the 1930s, a milk fortification scheme was implemented in the United States to fight rickets, considered to be a major public health problem. Other dairy products such as cheese are generally not fortified. Ready-to-eat breakfast snacks often contain vitamin D, similar to some brands of orange juice, yoghurt, margarine, and other foods. Both the United States and Canada require vitamin D-fortified formula for infants: 40–100 IU/100 kcal in the United States and 40–80 IU/100 kcal in Canada [28].

3.9 Calcium

Calcium is an essential element for the growth, activity, and maintenance of the human body [84]. Bone health is a major public concern. Each year, approximately

9 million people worldwide suffer from fractures due to osteoporosis [85]. Insufficient calcium and vitamin D absorption, an inadequate lifestyle, food choices, and genetics play an important role in the development of osteoporosis. Calcium absorption is controlled homeostatically by vitamin D regulation. Calcium deficiency can also lead to reduced blood clotting conditions, weak teeth, and some other symptoms [86]. People with lactose intolerance have decreased calcium intake due to absent dietary intake of calcium-dairy products [87], while the availability of calcium from nondairy sources is affected by the presence of the phytic acid, oxalic acid, and fiber [88]. Calcium is used for fortification in the form of calcium carbonate, calcium lysinate, and tricalcium phosphate in foods such as rice extrudates and noodles [89], tuna bone powder crackers [90], cookies [91], biscuits [92], yoghurt [84], etc.

The percentage of calcium absorbed depends on the total amount of elemental calcium consumed at one time; as the amount increases, the absorption percentage decreases. The absorption is highest in doses ≤500 mg. For example, someone who takes 1000 mg of calcium daily as supplements could divide the dose and take 500. The average dietary calcium intake for men over 1 year old is between 871 and 1266 mg/day, depending on the age group, while for females the interval is between 748 and 968 mg/day mg at two separate times during the day [29].

3.10 Selenium

Selenium (Se) is an essential micronutrient for both humans and animals. It forms an important component of glutathione peroxidase, a well-known antioxidant that counteracts cellular oxidative destruction. Furthermore, it plays an important role in catalyzing the production of the active thyroid hormone [93, 94], and it is required to improve human immunity and sperm motility [95]. Epidemiological studies have indicated that Se deficiency is positively correlated with the incidence of cancer [96]. The average daily recommended dose for an adult is 60 μg/day [97]. Selenium content in foods of plant origin varies with soil selenium content [98]. The most prevalent selenium fortification is that of the cooking salt, which significantly reduced the prevalence of Keshan disease in China [99], but there are other foods that can be fortified with selenium, such as yoghurt [100].

3.11 Fibers

Constipation has a negative impact on the quality of life and is defined as a difficult evacuation accompanied by discomfort and pain, while long-term constipation entails serious health problems. This is due to a decrease in dietary fiber consumption, which is directly proportional to excessive food processing [101].

Food fibers can be classified as soluble and insoluble fibers. Both types of fiber have many health benefits including maintaining intestinal and overall health integrity, lowering blood cholesterol levels, controlling blood sugar levels, and providing a non-caloric volume that can help weight loss by replacing caloric food components such as fat. According to the Dietary Guidelines for Americans, dietary fiber is consumed by most adults indicating that fortification of fiber foods might bring health benefits. Numerous studies [102–105] have shown that sausage fiber fortification (2–3 g/portion) can be achieved without any negative impact on sausage sensory quality. Alongside the increased fiber intake, products also come with other advantages such as fat replacement, increased water holding capacity, and improved oxidative stability when the fiber source is associated with phenolic antioxidants [103, 106], bran bread [107], and fiber-containing yoghurt [108].

3.12 Proteins

Proteins belong to a category of biologically active compounds that are essential to life through their specific action. They are the building blocks for the formation of tissues in the human body (cell walls, muscles, blood, hair, internal organs such as the heart and the brain, etc.), hormones, enzymes and antibodies, and replacement of waste cells. The essential amino acids are those that the body cannot produce by itself, and thus, they must be secured from food. Nonessential amino acids are amino acids that can be synthesized by the human body from essential acids or by cleavage of proteins [109]. The quality of plant proteins (vs. those from animal sources) has become a very debatable topic due to the consumption of plant products, which is increasingly promoted in the "world of nutrition." As the presence of essential amino acids in the body is of utmost importance, nutritionists recommend the right combination between the two sources (plant and animal). An advantage can be found in the addition of a vegetable matrix to meat preparation. This improves the mineral profile of the product and brings a supply of vitamins to the finished product, for example, mushroom sausages, meat products with soy protein isolate, and whey or wheat protein [110, 111], but also the fortification of plant products with plant proteins, for example, the fortification of wheat flour with cottonseed, chickpeas, amaranth, quinoa, and lentils [12].

3.13 Fatty acids

Essential fatty acids are important nutrients for the human body, especially for maintaining the health of the cardiovascular system. Several sources of information suggest that human beings have evolved on a diet that included a ratio of omega-6 to omega-3 fatty acids (EFA) of ~ 1, while western diets include a ratio of 15/1–16.7/1. Western diets exhibit omega-3 fatty acid deficiency, while they also include excessive amounts of omega-6 compared to the diet that human beings have evolved on and which set the foundation for their genetic patterns [112]. Products fortified with fatty acids in balanced proportions include meat, oil, jelly, various sauces, etc.

3.14 Multiple fortifications

Based on what is known about the prevalence of individual micronutrient deficiencies, there are generally multiple deficiencies of common micronutrients or in different population groups. Micronutrient deficiencies are more common in people on a diet low in foods of animal origin and therefore low in iron and bioavailable zinc intake, as well as calcium, retinol (vitamin A), vitamin B2 (riboflavin), vitamin B6, and vitamin B12 intakes. Commonly, poor diets, lacking fresh fruits and vegetables, yield deficiencies of vitamin C (ascorbic acid), β-carotene (provitamin A), and folic acid. Grain milling eliminates several nutrients, especially iron and zinc, various B complex vitamins (i.e., thiamine, riboflavin, and niacin), and folate. People on refined grain diets are at increased risk of deficiency in all these micronutrients. [4] Thus, fortifications are cumulated, such as vitamin A and iron, the vitamin B complex, calcium, and vitamin D, if micronutrients are separated. If fortification is achieved with the addition of different raw or auxiliary materials, the finished product will benefit from multiple fortifications, for example, the fortification of bread or sausages with mushrooms—this product will additionally have the nutritional value for protein and fiber content [110], while bakery products with the addition of nut paste come with a high intake of lipids, fibers, and minerals [12]. Most of the products fortified with unprocessed vegetable additions exhibit multiple fortifications.

4. Advanced applied technologies for the fortification of bakery products

The lentil flour was obtained by grinding the lentils, thus obtaining a fine consistency; then flour mixtures were created according to **Figure 1** [113].

Several aspects need to be taken into account throughout the technological process of obtaining fortified bakery products. This process of obtaining different types of fortified bread, including lentil bread, can be achieved through a single-phase process, since the amount (10–30%) of lentil flour is always introduced into the process by uniform distribution in the flour mass before adding the wet components (**Figure 1**) or through a two-phase process [113].

The use of compressed yeast as suspension in water facilitates the even distribution of yeast cells in the dough, thereby contributing to the improvement of product quality. As such, preparing the compressed yeast involves performing several operations such as suspending, filtering the suspension, and activating the yeast [114, 115]. The dosing aims at obtaining the dough with the optimal rheological properties and the appropriate composition of the product. The dosing of the raw materials used for dough preparation is performed taking into account their physical characteristics [113].

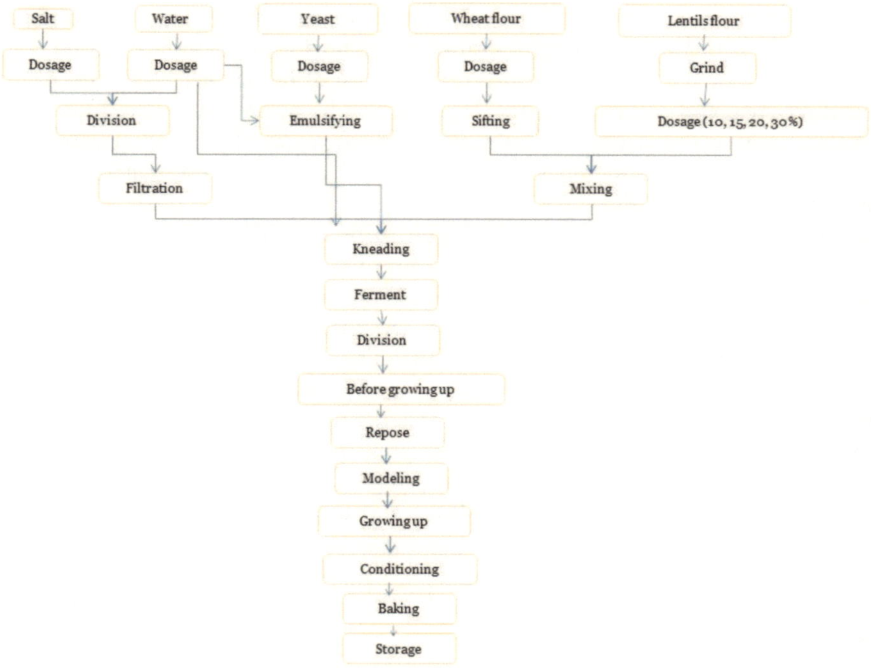

Figure 1.
The technological chart for the production of bread fortified with lentil flour, according to [113].

During the kneading process, an amount of air is included in the dough, required to meet its rheological attributes. The physical processes that take place in the dough at its kneading are the mechanical action during the kneading that ensures the water penetration into the flour mass and the temperature increase of the dough [116]. During the process of kneading itself, wet clumps of flour stick together under the influence of the mechanical kneading action, surface water penetrates deep, proteins become hydrated, the amount of bound water increases, and the dough increases its consistency and gradually acquires its elastic properties [117].

The fermenting operation is aimed at obtaining a risen dough that would produce suitable well-risen products with the right volume, as well as a porous and elastic core. Therefore, during this operation, various dough substances are accumulated in the dough, which impart the taste and aroma of bakery products. During dough fermentation, the yeast multiplication process continues, the predominant process being alcoholic fermentation, which releases carbon dioxide, ethyl alcohol, and a small amount of heat [116].

A well-risen product can only be obtained if the dough forms large amounts of gas. Ultimate leavening is the operation where the leavening of the dough reaches the maximum. The gases formed affect the gluten skeleton and thus the porosity of the dough by increasing or decreasing mesh size [116]. The biochemical processes formed during fermentation are hydrolysis of starch and gluten amylolysis or proteolysis, respectively. The microbiological processes refer to the multiplication of yeasts and their fermentative activity, as well as that of lactic bacteria that make up dough microbiota. Colloidal processes are the continuation of the gluten-forming process and gluten peptization [115].

Some of the gas present in the dough is lost during modeling, and as a result, the spongy structure of the dough is largely destroyed, the internal surface is reduced, and its specific weight increased [115].

A significant part of the carbon dioxide accumulated in the dough is eliminated during the division and modeling operations. Therefore, the dough must be subjected to a new fermentation for the restoration of its porous structure. As such, the products rise and the volume develops [116]. The main goal of the final leavening of the dough is the rise by the accumulation of carbon dioxide that is formed during yeast alcoholic fermentation and conditions the volume and structure of the porosity of products [115].

Colloidal processes condition the dough to be transformed into the inner crumb and modify the water binding state so that the crumb, although exhibiting a higher moisture than the initial humidity of the dough, appears dry due to thermal hydrophilic modifications. During coagulation, protein clotting and the gelatinization of starch occur [116]. The protein coagulation process starts at temperatures slightly above 50° C and proceeds at a maximum speed of 60–70° C, thus accelerating to continue the heating of the dough. After coagulation, proteins become more easily attackable by enzymes, both digestive enzymes and the proteolytic ones in the dough. The gelatinization of the starch takes place in two stages: granule swelling and gelatinization itself. Swelling occurs due to the penetration of water groups within the starch granule that distances the protein chains. This is achieved by increasing the kinetic energy of water molecules by heating the dough. The gelatinization is accomplished by breaking the existing H2 bonds between the amylose and amylopectin chains and dispersing these chains between the molecules of water previously penetrated into the granule. Biochemical processes formed during baking are amylolysis and proteolysis. Microbial processes during baking include alcoholic fermentation and lactic fermentation [117].

5. Results obtained in food fortification

After lentil flour was employed in 10%, 15%, 20%, and 30% proportions for the fortification of bread, the increase in mineral elements was found to be up to 4 times the control sample, and protein content increased from 6.55% to 10.50% and fibers from 0.6% to 4.7%, while the ratio of ω-6 and ω-3 fatty acids decreased from 16.66 to 7.86 [113].

For 10%, 20%, and 30% flax fiber fortified biscuits, the results showed that the total content of polyunsaturated fatty acids increased significantly from 10.50% to 20.50%, while the total content of saturated fatty acids decreased significantly from 48.22% to 40.20%, showing that the omega-3 content of fortified flax seed biscuits was significantly higher than biscuits lacking flax seeds, while the ω-6/ω-3 dropped significantly from 8.50 to 1.00. Furthermore, the calcium content increased significantly from 25.6 mg/100 g for the control sample to 168.3, 175.0, and 179.1 mg/100 g, respectively [92].

Examples may continue for different products: mushroom sausages [110], mushroom powder pastes [118], marc bread [119], chestnut flour bread [120], spirulina paste [121], inulin- and lactobacilli-fortified juice [122], tortillas fortified with iron and folic acid [123, 124], bread fortified with zinc and iron [19], etc.

6. Conclusion

The benefits of food fortification positively impact the entire life cycle of mankind. Thus, food fortification may be one of the most effective ways to overcome malnutrition and various diseases, especially in children and pregnant women and their children, preventing the birth of intellectually impaired children with malformations or deficiencies. Efforts to fortify food must be integrated in the context of each country's public health and nutrition plans and as part of a comprehensive strategy to include different micronutrients in staple foods to meet the social, physiological, and economic goals of millions of people worldwide.

Author details

Vlaic Romina Alina[1], Mureşan Crina Carmen[1*], Muste Sevastita[1], Mureşan Andruța[1], Muresan Vlad[1], Suharoschi Ramona[2], Petruț Georgiana[1] and Mihai Mihaela[2]

1 Department of Food Engineering, Faculty of Food Science and Technology, University of Agricultural Sciences and Veterinary Medicine of Cluj-Napoca, Cluj-Napoca, Romania

2 Department of Food Science, Faculty of Food Science and Technology, University of Agricultural Sciences and Veterinary Medicine of Cluj-Napoca, Cluj-Napoca, Romania

*Address all correspondence to: crina.muresan@usamvcluj.ro

References

[1] Codex Alimentarius Commission. General Principles for the Addition of Essential Nutrients to Foods CAC/GL 09-1987 (amended 1989, 1991). Rome, Joint FAO/WHO Food Standards Programme, Codex Alimentarius Commission, 1987 (http://www.codexalimentarius.net/download/standards/299/CXG_009e.pdf [Accessed: Oct 1, 2018])

[2] Burgi H, Supersaxo Z, Selz B. Iodine deficiency diseases in Switzerland one hundred years after Theodor Kocher's survey: A historical review with some new goitre prevalence data. Acta Endocrinologica. 1990;**123**:577-590. DOI: 10.1530/acta.0.1230577

[3] Marine D, Kimball OP. Prevention of simple goiter in man. Archives of Internal Medicine. 1920;**25**:661-672. DOI: 10.1001/archinte.1920.00090350088005

[4] Lindsay A, de Benoist B, Dary O, Hurrell R. Guidelines on Food Fortification with Micronutrients. Geneva, Switzerland: WHO Library; 2006. http://www.who.int/entity/nutrition/publications/guide_food_fortification_micronutrients.pdf. [Accessed: Oct 1, 2018]

[5] Mozaffarian D, Rosenberg I, Uauy R. History of modern nutrition science—Implications for current research, dietary guidelines, and food policy. BMJ. 2018;**361**(k2392):1-6. DOI: 10.1136/bmj.k2392

[6] Darnton-Hill I, Nalubola R. Fortification strategies to meet micronutrient needs: Successes and failures. The Proceedings of the Nutrition Society. 2002;**61**:231-241. DOI: 10.1079/PNS2002150

[7] Samoraj M, Tuhy Ł, Dmytryk A, Chojnacka K. Biofortification of food with trace elements. In: Recent Advances in Trace Elements. Wiley Online Library. 2018;**21**:443-456. DOI: 10.1002/9781119133780.ch21

[8] Fageria NK, Moraes MF, Ferreira EP, Knupp AM. Biofortification of trace elements in food crops for human health. Communications in Soil Science and Plant Analysis. 2012; **43**(3):556-570. DOI: 10.1080/00103624.2012.639431

[9] Salem HM, El–Gizawy NK. Importance of micronutrients and its application methods for improving maize (*Zea mays* L.) yield grown in clayey soil. Chemist-Analyst. 2012; **12**(7):954-959. DOI: 10.5829/idosi.aejaes.2012.12.07.1759

[10] Shi R, Zhang Y, Chen X, Sun Q, Zhang F, Römheld V, et al. Influence of long-term nitrogen fertilization on micronutrient density in grain of winter wheat (*Triticum aestivum* L.). Journal of Cereal Science. 2010;**51**(1):165-170. DOI: 10.1016/j.jcs.2009.11.008

[11] Kumar N, Kumar RJ, Kumar CV, Jain S, Rajesh. Backcross breeding for enhancing minerals (iron and zinc) content in rice (*Oryza sativa* L.). International Journal of Current Microbiology and Applied Sciences. 2018;7(5):3593-3603. DOI: 10.20546/ijcmas.2018.705.415

[12] Păucean A. Tendințe modern privind creşterea valorii nutritive a făinii de grâu şi a produselor de panificație. Cluj Napoca: Editura Mega; 2017

[13] Velu G, Ortiz–Monasterio I, Cakmak I, Hao Y, Singh RP. Biofortification strategies to increase grain zinc and iron concentrations in wheat. Journal of Cereal Science. 2014; **59**(3):365-372. DOI: 10.1016/j.jcs.2013.09.001

[14] Osendarp SJM, Martinez H, Garrett GS, Neufeld LM, De-Regil LM, Vossenaar M, et al. Large-scale food fortification and biofortification in low- and middle-income countries: A review of programs, trends, challenges, and evidence gaps. Food and Nutrition Bulletin. 2018;**39**(2):315-331. DOI: 10.1177/0379572118774229

[15] Saha S, Chakraborty M, Padhan D, Saha B, Murmu S, Batabyal K, et al. Agronomic biofortification of zinc in rice: Influence of cultivars and zinc application methods on grain yield and zinc bioavailability. Field Crops Research. 2017;**210**:52-60. DOI: 10.1016/j.fcr.2017.05.023

[16] Lidon FC, Oliveira K, Ribeiro MM, Pelica J, Pataco I, Ramalho JC, et al. Selenium biofortification of rice grains and implications on macronutrients quality. Journal of Cereal Science. 2018; **81**:22-29. DOI: 10.1016/j.jcs.2018.03.010

[17] Wu J, Salisbury C, Graham R, Lyons G, Fenech M. Increased consumption of wheat biofortified with selenium does not modify biomarkers of cancer risk, oxidative stress, or immune function in healthy Australian males. Environmental and Molecular Mutagenesis. 2009;**50**:489-501. DOI: 10.1002/em.20490

[18] Naz S, Yousaf B, Tahir MA, Qadir A, Yousaf A. Iron and zinc bio-fortification strategies in wheat crop by exogenous application of micronutrients. Food Science and Quality Management. 2015; **35**:49-55

[19] Ciccolini V, Pellegrino E, Coccina A, Fiaschi AI, Cerretani D, Sgherri C, et al. Biofortification with iron and zinc improves nutritional and nutraceutical properties of common wheat flour and bread. Journal of Agricultural and Food Chemistry. 2017;**65**(27):5443-5452. DOI: 10.1021/acs.jafc.7b01176

[20] The World Health Report 2002. Reducing Risks, Promoting Healthy Life: Overview. Geneva: World Health Organization; 2002. Available from: http://www.who.int/whr/2002/en/

[21] Liu OR, Molina R, Wilson M, Halpern BS. Global opportunities for mariculture development to promote human nutrition. PeerJ. 2018;**6**:e4733. DOI: 10.7717/peerj.4733

[22] Joosten E. Iron deficiency anemia in older adults: A review. Geriatrics & Gerontology International. 2017;**18**(3): 373-379. DOI: 10.1111/ggi.13194

[23] Erdman JW, MacDonald AI, Steven HZ. Present Knowledge in Nutrition. 10th ed. Vol. 33. Hong Kong: Wiley-Blackwell; 2012. pp. 506-520. DOI: 10.1002/9781119946045

[24] Doran L. Echocardiography Audit on Patients with Hereditary Haemochromatosis. Masters thesis. DIT; 2016

[25] Drakesmith H, Prentice AM. Hepcidin and the iron-infection axis. Science. 2012;**338**:768-772. DOI: 10.1126/science.1224577

[26] Trumbo P, Yates AA, Schlicker S, Poos M. Dietary reference intakes. Journal of the American Dietetic Association. 2001;**101**(3):294-301. DOI: 10.1016/s0002-8223(01)00078-5

[27] Institute of Medicine (US) Standing Committee on the Scientific Evaluation of Dietary Reference Intakes. Dietary Reference Intakes for Thiamin, Riboflavin, Niacin, Vitamin B6 , Folate, Vitamin B12, Pantothenic Acid, Biotin, and Choline. Washington, DC, USA: National Academy of Sciences; 1998

[28] Institute of Medicine. Food and Nutrition Board. Dietary Reference Intakes for Calcium and Vitamin D. Washington, DC: National Academy Press; 2010

[29] Committee to Review Dietary Reference Intakes for Vitamin D and Calcium, Food and Nutrition Board, Institute of Medicine. Dietary Reference Intakes for Calcium and Vitamin D. Washington, DC: National Academy Press; 2010

[30] Institute of Medicine, Food and Nutrition Board. Dietary Reference Intakes. Vitamin C, Vitamin E, Selenium, and Carotenoids. Washington DC: National Academy Press; 2000

[31] Institute of Medicine. Dietary Reference Intakes for Energy, Carbohydrate, Fiber, Fat, Fatty Acids, Cholesterol, Protein, and Amino Acids. National Academy Press. Food and Nutrition Board, Institute of Medicine, National Academies. 2005

[32] Baker RD, Greer FR. Diagnosis and prevention of iron deficiency and iron-deficiency anemia in infants and young children (0–3 years of age). Pediatrics. 2010;**126**:1040-1050. DOI: 10.1542/peds.2010-2576

[33] Whittaker P, Tufaro PR, Rader JI. Iron and folate in fortified cereals. Journal of the American College of Nutrition. 2001;**20**:247-254. DOI: 10.1080/07315724.2001.10719039

[34] Andrew TG, Huchzermeyer KDA, Mbeha BC, Nengu SM, Arthur JR. Scientific opinion on dietary reference values for iron. EFSA Journal. 2015; **13**(10):4254. DOI: 10.2903/j.3efsa.2015.4254

[35] Cogswell ME, Parvanta I, Ickes L, Yip R, Brittenham GM. Iron supplementation during pregnancy, anemia, and birth weight: A randomized controlled trial. The American Journal of Clinical Nutrition. 2003;**78**(4):773-781. DOI: 10.1093/ajcn/78.4.773

[36] Thuy PV, Berger J, Davidsson L, Khan NC, Lam NT, Cook JD, et al. Regular consumption of NaFeEDTA-fortified fish sauce improves iron status and reduces the prevalence of anemia in anemic Vietnamese women. The American Journal of Clinical Nutrition. 2003;**78**(2):284-290. DOI: 10.1093/ajcn/78.2.284

[37] Mannar V, Boy Gallego E. Iron fortification: Country level experiences and lessons learned. The Journal of Nutrition. 2002;**132**:856S-858S. DOI: 10.1093/jn/132.4.856S

[38] Ballot DE, MacPhail AP, Bothwell TH, Gillooly M, Mayet FG. Fortification of curry powder with NaFe(111)EDTA in an iron-deficient population: Report of a controlled iron-fortification trial. The American Journal of Clinical Nutrition. 1989;**49**(1):162-169. DOI: 10.1093/ajcn/49.1.162

[39] Johnson EJ, Russell RM. Beta-carotene. In: Coates PM, Betz JM, Blackman MR, et al., editors. Encyclopedia of Dietary Supplements. 2nd ed. London and New York: Informa Healthcare; 2010. pp. 115-120

[40] Ross CA. Vitamin A. In: Coates PM, Betz JM, Blackman MR, et al., editors. Encyclopedia of Dietary Supplements. 2nd ed. London and New York: Informa Healthcare; 2010. pp. 778-791

[41] Ross A. Vitamin A and carotenoids. In: Shils M, Shike M, Ross A, Caballero B, Cousins R, editors. Modern Nutrition in Health and Disease. 10th ed. Baltimore, MD: Lippincott Williams & Wilkins; 2006. pp. 351-375

[42] Global Prevalence of Vitamin A Deficiency. Micronutrient Deficiency Information System working paper No. 2. Geneva, World Health Organization. 1995. Available from: http://www.who.int/nutrition/publications/micronutrients/iodine_deficiency/54015_mdis_workingpaper1/en/

[43] Indicators for Assessing Vitamin A Deficiency and their Application in Monitoring and Evaluating Intervention Programmes. Geneva: World Health Organization; 1996. Available from: http://www.who.int/nutrition/publications/micronutrients/vitamin_a_deficiency/WHO_NUT_96.10/en/

[44] Beaton GH, Martorell R, Aronson KJ, Edmonston B, McCabe G, Ross AC, et al. Effectiveness of Vitamin a Supplementation in the Control of Young Child Morbidity and Mortality in Developing Countries. Geneva: Administrative Committee on Coordination—Sub-Committee on Nutrition; 1992. Available from: https://www.unscn.org/web/archives_resources/files/Policy_paper_No_13.pdf

[45] Solon FS, Solon MS, Mehansho H, West KP, Sarol J, Perfecto C, et al. Evaluation of the effect of vitamin A-fortified margarine on the vitamin A status of preschool Filipino children. European Journal of Clinical Nutrition. 1996;**50**:720-723

[46] National Research Council. Committee to Assess the Health Implications of Perchlorate Ingestion. Health Implications of Perchlorate Ingestion. Washington, DC: The National Academies Press; 2005

[47] World Health Organization. United Nations Children's Fund & International Council for the Control of Iodine Deficiency Disorders. Assessment of Iodine Deficiency Disorders and Monitoring their Elimination. 3rd ed. Geneva, Switzerland: WHO; 2007

[48] Zimmermann MB. Iodine deficiency. Endocrine Reviews. 2009; **30**(4):376-408. DOI: 10.1210/er.2009-0011

[49] Zimmermann MB, Jooste PL, Pandav CS. Iodine-deficiency disorders. Lancet. 2008;**372**(9645):1251-1262. DOI: 10.1016/S0140-6736(08)61005-3

[50] Murray CW, Egan SK, Kim H, Beru N, Bolger PM. US Food and Drug Administration's total diet study: Dietary intake of perchlorate and iodine. Journal of Exposure Science & Environmental Epidemiology. 2008; **18**(6):571-580. DOI: 10.1038/sj.jes.7500648

[51] Bailey LB, Gregory JF 3rd. Folate. In: Bowman B, Russell R, editors. Present Knowledge in Nutrition. Washington, DC: International Life Sciences Institute; 2006. pp. 278-301

[52] Carmel R. Folic acid. In: Shils M, Shike M, Ross A, Caballero B, Cousins R, editors. Modern Nutrition in Health and Disease. Baltimore, MD: Lippincott Williams & Wilkins; 2005. pp. 470-481

[53] Yetley EA, Pfeiffer CM, Phinney KW, Fazili Z, Lacher DA, Bailey RL, et al. Biomarkers of folate status in NHANES: A roundtable summary. The American Journal of Clinical Nutrition. 2011;**94**(1):303-312. DOI: 10.3945/ajcn.111.013011

[54] Colapinto CK, O'Connor DL, Dubois L, Tremblay MS. Folic acid supplement use is the most significant predictor of folate concentrations in Canadian women of childbearing age. Applied Physiology, Nutrition, and Metabolism. 2012;**37**(2):284-292. DOI: 10.1139/h11-161

[55] U.S. Department of Agriculture Agricultural Research Service. USDA National Nutrient Database for Standard Reference, Release 252012. DOI: 10.13140/RG.2.1.1551.1284

[56] U.S. Food and Drug Administration. Food standards: Amendment of standards of identity for enriched grain products to require addition of folic acid. Federal Register. 1996;**61**(44)

[57] Choumenkovitch SF, Selhub J, Wilson PWF, Rader JI, Rosenberg IH, Jacques PF. Folic acid intake from fortification in United States exceeds predictions. The Journal of Nutrition. 2002;**132**(9):2792-2798. DOI: 10.1093/jn/132.9.2792

[58] Food and Drug Administration. Final rule: Food additives permitted for direct addition to food for human consumption; folic acid. Federal Register. 2016;**81**(73):22176-22183

[59] Wilson RD, Désilets V, Wyatt P, Langlois S, Gagnon A, et al. Pre-conceptional vitamin/folic acid supplementation 2007: The use of folic acid in combination with a multivitamin supplement for the prevention of neural tube defects and other congenital anomalies. Journal of Obstetrics and Gynaecology Canada. 2007;**29**(12):1003-1013. DOI: 10.1016/s1701-2163(16)32685-8

[60] Centers for Disease Control and Prevention. CDC grand rounds: Additional opportunities to prevent neural tube defects with folic acid fortification. MMWR. Morbidity and Mortality Weekly Report. 2010;**59**(31):980-984

[61] Moll R, Davis B. Iron, vitamin B12 and folate. Medicine. 2017;**45**(4):198-203. DOI: 10.1016/j.mpmed.2017.01.007

[62] Carmel R. How I treat cobalamin (vitamin B12) deficiency. Blood. 2008;**112**:2214-2221. DOI: 10.1182/blood-2008-03-040253

[63] Vitamin and Mineral Requirements in Human Nutrition. Report of a Joint FAO/WHO Expert Consultation on Human Vitamin and Mineral Requirements. Bangkok, Thailand, 21–30 September 1998. 2nd ed. Geneva: World Health Organization; 2004. Available from: http://apps.who.int/iris/bitstream/handle/10665/42716/9241546123.pdf?sequence=1

[64] Andrès E, Federici L, Affenberger S, Vidal-Alaball J, Loukili NH, Zimmer J, et al. B12 deficiency: A look beyond pernicious anemia. The Journal of Family Practice. 2007;**56**:537-542

[65] U.S. Department of Agriculture, Agricultural Research Service. USDA National Nutrient Database for Standard Reference, Release 24. . Nutrient Data Laboratory Home Page. 2011

[66] Ezzati M, Riboli E. Behavioral and dietary risk factors for noncommunicable diseases. The New England Journal of Medicine. 2013;**369**:954-964. DOI: 10.1056/NEJMra1203528

[67] Institute of Medicine (US) Standing Committee on the Scientific Evaluation of Dietary Reference Intakes. The B Vitamins and Choline: Overview and Methods. Washington, DC, USA: National Academies Press; 1998

[68] Ortigoza-Escobar JD, Alfadhel M, Molero-Luis M, Darin N, Spiegel R, de Coo IF. Thiamine deficiency in childhood with attention to genetic causes: Survival and outcome predictors. Annals of Neurology. 2017;**82**(3):317-330. DOI: 10.1002/ana.24998

[69] European Food Safety Authority (EFSA). Dietary reference values for riboflavin. EFSA Journal. 2017;**18**:4919

[70] European Food Safety Authority (EFSA). Scientific opinion on dietary reference values for niacin. EFSA Journal. 2014;**12**:3759

[71] European Food Safety Authority (EFSA). Dietary reference values for vitamin B6. EFSA Journal. 2016;**14**:4487

[72] Rosenberg IH, Abrams SA, Beecher GR, et al. Dietary reference intakes: Guiding principles for nutrition labeling

and fortification. Nutrition Reviews. 2004;**62**:73-79

[73] Swaminathan S, Thomas T, Kurpad AV. B-vitamin interventions for women and children in low income populations. Current Opinion in Clinical Nutrition and Metabolic Care. 2015;**18**:295-306. DOI: 10.1097/MCO.0000000000000166

[74] Scholey A. Vitamin B1 deficiency; vitamins in human nutrition—Social aspects; vitamins—Health aspects; vitamin B deficiency; health—Physiological aspects. The Australian Journal of Pharmacy. 2018;**1175**:58-64

[75] Whitfield KC, Karakochuk CD, Kroeun H, Sokhoing L, Chan BB, Borath M, et al. Household consumption of thiamin-fortified fish sauce increases erythrocyte thiamin concentrations among rural Cambodian women and their children younger than 5 years of age: A randomized controlled efficacy trial. The Journal of Pediatrics. 2017;**181**: 242-247. DOI: 10.1016/j.jpeds.2016.10.066

[76] FAO. FAOSTAT. Food Balance Sheet Cambodia. 2011. Available from: http://www.fao.org/faostat/en/#country/115

[77] Reidy KC, Bailey RL, Deming DM, O'Neill L, Carr BT, Lesniauskas R, et al. Food consumption patterns and micronutrient density of complementary foods consumed by infants fed commercially prepared baby foods. Nutrition Today. 2018;**53**(2): 68-78. DOI: 10.1097/NT.0000000000000265

[78] Thakur NS, Thakur N, Hamid, Kumar P, Thakur A. Formulation and optimization of vitamin C rich ready-to-serve juice based beverage from wild aonla fruits and its quality evaluation during storage. Journal of Pharmacognosy and Phytochemistry. 2018;**7**(1):1796-1802

[79] Mohammadi T, Azandeh S, bijan Nejad D, Shamseddini lori N. Effect of vitamin C on bioavailability and differentiation potential of human umbilical cord Wharton's jelly-derived mesenchymal stem cells. Qom University of Medical Sciences Journal. 2018;**11**(11):1-11

[80] Wang AH, Still C. Old world meets modern: A case report of scurvy. Nutrition in Clinical Practice. 2007;**22**: 445-448. DOI: 10.1177/0115426507022004445

[81] Pilz S, März W, Cashman KD, Kiely ME, Whiting SJ, Holick MF, et al. Rationale and plan for vitamin D food fortification: A review and guidance paper. Frontiers in Endocrinology. 2018; **9**:1-16. DOI: 10.3389/fendo.2018.00373

[82] Singh GM, Micha R, Khatibzadeh S, Shi P, Lim S, Rews KG, et al. Global, regional, and national consumption of sugar-sweetened beverages, fruit juices, and milk: A systematic assessment of beverage intake in 187 countries. PLoS One. 2015;**10**(8):1-20. DOI: 10.1371/journal.pone.0124845

[83] Calvo MS, Whiting SJ, Barton CN. Vitamin D fortification in the United States and Canada: Current status and data needs. The American Journal of Clinical Nutrition. 2004;**80**:1710S-1716S. DOI: 10.1093/ajcn/80.6.1710S

[84] El-Shibiny S, El-Gawad M, Assem FM, El-Sayed SM. The use of nano-sized eggshell powder for calcium fortification of cow's and buffalo's milk yogurts. Acta Scientiarum Polonorum. Technologia Alimentaria. 2018;**17**(1): 37-49. DOI: 10.17306/J.AFS.0541

[85] Curtis EM, Moon RJ, Dennison EM, Harvey NC, Cooper C. Recent advances in the pathogenesis and treatment of osteoporosis. Clinical Medicine. 2015; **15**(6):92-96. DOI: 10.7861/clinmedicine.15-6-s92

[86] Meschino J. Calcium: Requirements, bioavailable forms, physiology and clinical aspects. Dynamic Chiropractic. 2002;**20**(18):1-12

[87] Heaney RP, Rafferty K, Dowell MS, Bierman J. Calcium fortification systems differ in bioavailability. Journal of the American Dietetic Association. 2005; **105**(5):807-809. DOI: 10.1016/j.jada.2005.02.012

[88] Fulgoni VL, Keast DR, Auestad N, Quann EE. Nutrients from dairy foods are difficult to replace in diets of Americans: Food pattern modeling and an analyses of the National Health and Nutrition Examination Survey 2003–2006. Nutrition Research. 2011;**31**(10): 759-765. DOI: 10.1016/j.nutres.2011.09.017

[89] Janve M, Singhal RS. Fortification of puffed rice extrudates and rice noodles with different calcium salts: Physicochemical properties and calcium bioaccessibility. Lebensmittel-Wissenschaft & Technologie. 2018;**97**: 67-75. DOI: 10.1016/j.lwt.2018.06.030

[90] Benjakul S, Karnjanapratum S. Characteristics and nutritional value of whole wheat cracker fortified with tuna bone bio-calcium powder. Food Chemistry. 2018;**259**:181-187. DOI: 10.1016/j.foodchem.2018.03.124

[91] Agrahar-Murugkar D, Dwivedi S, Dixit-Bajpai P, Kumar M. Effect of natural fortification with calcium and protein rich ingredients on texture, nutritional quality and sensory acceptance of cookies. Nutrition & Food Science. 2018. DOI: 10.1108/nfs-02-2018-0041

[92] Elshehy H, Agamy N, Ismail H. Effect of fortification of biscuits with flaxseed on omega 3 and calcium content of the products. Journal of High Institute of Public Health. 2018;**48**(2): 58-66

[93] Wojciechowska-Durczynska K, Lewinski A. Search for relevant indications for selenium supplementation in thyroid diseases. Neuroendocrinology Letters. 2017;**38**: 237-241

[94] Stuss M, Michalska-Kasiczak M, Sewerynek E. The role of selenium in thyroid gland pathophysiology. Endokrynologia Polska. 2017;**68**: 440-454. DOI: 10.5603/EP.2017.0051

[95] Gill H, Walker G. Selenium, immune function and resistance to viral infections. Nutrition and Dietetics. 2008;**65**:41-47. DOI: 10.1111/j.1747-0080.2008.00260.x

[96] Rayman MP. Selenium in cancer prevention: A review of the evidence and mechanism of action. The Proceedings of the Nutrition Society. 2005;**64**:527-542. DOI: 10.1079/PNS2005467

[97] Gao J, Liu Y, Huang Y, Lin ZQ, Banuelos GS, Lam MHW, et al. Daily selenium intake in a moderate selenium deficiency area of Suzhou, China. Food Chemistry. 2011;**126**:1088-1093. DOI: 10.1016/j.foodchem.2010.11.137

[98] Fox TE, Fairweather-Tait S. Selenium. In: Hurrell RF, editor. The Mineral Fortification of Foods. Leatherhead, Surrey: Leatherhead Publishing; 1999. pp. 112-153

[99] Cheng YY, Qian PC. The effect of selenium-fortified table salt in the prevention of Keshan disease on a population of 1.05 million. Biomedical and Environmental Sciences. 1990;**3**: 422-428

[100] Alsuhaibani AMA. Functional role of selenium-fortified yogurt against aflatoxin-contaminated nuts in rats. Agriculture & Food Security. 2018;**7**(1). DOI: 10.1186/s40066-018-0171-7

[101] Dahl WJ, Whiting SJ, Healey A, Zello GA, Hildebrandt SL. Increased stool frequency occurs when finely processed pea hull fiber is added to usual foods consumed by elderly residents in long-term care. Journal of the American Dietetic Association. 2003;**103**(9):1199-1202. DOI: 10.1016/s0002-8223(03)00977-5

[102] Besbes S, Attia H, Deroanne C, Makni S, Blecker C. Partial replacement of meat by pea fiber and wheat fiber: Effect on the chemical composition, cooking characteristics and sensory properties of beef burgers. Journal of Food Quality. 2008;**31**(4):480-489. DOI: 10.1111/j.1745-4557.2008.00213.x

[103] Choi Y, Choi J, Han D, Kim H, Lee M, Lee E, et al. Effects of rice bran fiber on quality of low-fat tteokgalbi. Food Science and Biotechnology. 2008;**17**(5):959-964

[104] Salazar P, Garcia ML, Selgas MD. Short-chain fructooligosaccharides as potential functional ingredient in dry fermented sausages with different fat levels. International Journal of Food Science and Technology. 2009;**44**(6):1100-1107. DOI: 10.1111/j.1365-2621.2009.01923.x

[105] Yilmaz I, Gecgel U. Effect of inulin addition on physico-chemical and sensory characteristics of meatballs. Journal of Food Science and Technology-Mysore. 2009;**46**(5):473-476

[106] Sayago-Ayerdi SG, Brenes A, Goni I. Effect of grape antioxidant dietary fiber on the lipid oxidation of raw and cooked chicken hamburgers. LWT-Food Science and Technology. 2009;**42**(5):971-976. DOI: 10.1016/j.lwt.2008.12.006

[107] Pop F. Improvement of the nutritional value of pasta by the addition of wheat bran. Journal of Agroalimentary Processes and Technologies. 2017;**23**(4):245-249

[108] Mudgil D, Barak S, Khatkar BS. Development of functional yoghurt via soluble fiber fortification utilizing enzymatically hydrolyzed guar gum. Food Bioscience. 2006;**14**:28-33. DOI: 10.1016/j.fbio.2016.02.003

[109] Fürst P, Stehle P. What are the essential elements needed for the determination of amino acid requirements in humans? The Journal of Nutrition. 2004;**134**(6):1558S-1565S. DOI: 10.1093/jn/134.6.1558S

[110] Nagy M, Semeniuc CA, Socaci SA, Pop CR, Rotar AM, Sălăgean CD, et al. Utilization of brewer's spent grain and mushrooms in fortification of smoked sausages. Food Science and Technology. 2017;**37**(2):315-320. DOI: 10.1590/1678-457x.23816

[111] Zhang W, Xiao S, Samaraweera H, Lee EJ, Ahn DU. Improving functional value of meat products. Meat Science. 2010;**86**(1):15-31. DOI: 10.1016/j.meatsci.2010.04.018

[112] Simopoulos AP. Dossier: Polyunsaturated fatty acids in biology and diseases. The importance of the ratio of omega-6/omega-3 essential fatty acids. Biomedicine & Pharmacotherapy. 2002;**56**:365-379

[113] Moldovan OP. Study quality parameters obtained from bakers flour fortified with legumes. PhD Thesis. USAMV-CN; 2016

[114] Modoran C. Tehnologia produselor făinoase. Cluj-Napoca: Editura Academic Press; 2003

[115] Bordei D. Tehnologia modernă a panificației. București: Editura AGIR; 2004

[116] Păucean A, Man S. Tehnologia produselor vegetale. Cluj Napoca: Editura Academic Press; 2015

[117] Modoran C. Tehnologia morăritului și a panificației. Cluj-Napoca: Editura Risoprint; 2007

[118] Mureșan CC, Fărcaș A, Man S, Suharoschi R, Vlaic RA. Obtaining a functional product through the exploitation of mushroom flour in pasta. Bulletin UASVM Food Science and Technology. 2017;**74**:17-22. DOI: 10.15835/buasvmcn-fst:12641

[119] Fărcaş AC, Socaci SA, Dulf FV, Tofană M, Mudura E, Diaconeasa Z. Volatile profile, fatty acids composition and total phenolics content of brewers' spent grain by-product with potential use in the development of new functional foods. Journal of Cereal Science. 2015;**64**:34-42. DOI: 10.1016/j.jcs.2015.04.003

[120] Man S, Păucean A, Muste S, Mureşan C, Frâncu AV. Chestnut flour addition influence on bread quality. Journal of Agroalimentary Processes and Technologies. 2012;**18**(2):150-154

[121] Muresan C, Pop A, Man S, Socaci S, Fărcas A, Nagy M, et al. The influence of different proportions of spirulina (*Arthrospira plantensis*) on the quality of pasta. Journal of Agroalimentary Processes and Technologies. 2016;**22**(1): 24-27

[122] Khezri S, Mahmoudi R, Dehghan P. Fig juice fortified with inulin and *Lactobacillus delbrueckii*: A promising functional food. Applied Food Biotechnology. 2018;**5**(2):97-106. DOI: https://doi.org/10.22037/afb.v5i2.19844

[123] Martínez-Bustos F, Cruz Sánchez A, Ortega-Martínez A, del C, Aguilar-Palazuelos E. Study of the functionality of nixtamalized maize flours and tortillas added with microcapsules of ferrous fumarate and folic acid. Cereal Chemistry. 2018;**95**:699-707. DOI: 10.1002/cche.10084

[124] Popel S, Draganova E, Parşacova L, Colesnicenco A, Evtodienco L, Cropotova J. Fortifierea făinii de grâu cu suplimente de vitamine şi minerale. In: Papers of the Sibiu Alma Mater University Conference, FLftfa Editiom 24–26 March 2011. Sibiu

Chapter 7

Fibers: Healthy Component in Whole Wheat and Rye Flours

María Laura Gómez Castro, Claudia Cecilia Larregain, Ethel Noemi Coscarello and Roberto Jorge Aguerre

Abstract

The demand for foods made with whole grain ingredients that improve health continues to grow. The health benefits of whole grain cereal are well recognized and are attributed to the presence of dietary fiber and phytochemicals. Consumption of whole grain cereals has been associated with protection against cardiovascular disease and type 2 diabetes. The wheat and rye whole grain flours have a total fiber content that gives a healthy contribution for the daily feeding of the population. After applying a malting process, the composition of fibers is modified by increasing the fraction of soluble fibers. These fibers are prebiotic and improve flour functionality. In this chapter, we will study the properties of whole malted flour of wheat and rye and how it benefits health.

Keywords: functional flour, fibers in cereal, malting process, prebiotic, arabinoxylans

1. Introduction

Food and Agriculture Organization (FAO) of the United Nations and World Health Organization (WHO) expert's meeting in human nutrition agreed that carbohydrate intake should be at least 55 parts in 100 of total calories [1]. Caloric intake should be mainly complex carbohydrate available with low glycemic index. This recommendation emphasizes the consumption of foods that meet these properties. The term "fiber" means carbohydrate polymers, which are neither digested nor absorbed in the human small intestine. The properties of dietary fiber, as the retention capacity of water and oil, have beneficial results on food products by improving their organoleptic characteristics and their physiological effects on the human organism. The quality of the fibers varies according to the type of cereals and according to the processes that are carried out to obtain their flour. In previous works carried out in our laboratory and in other research centers, on rye malt, this procedure generates an increase of soluble fibers, maintaining total fiber content.

The flour made form rye has greater fiber content than the flour wheat. The properties of dietary fiber have beneficial physiological effects on the human organism. Some components of soluble fiber are called prebiotics, defined as food ingredients that positively affect the host.

Prebiotics are defined as nondigestible substances. The dietary fiber maintains biological activity within the human organism by selective stimulation of growth of beneficial microorganisms either naturally present or therapeutically introduced to intestine. The intestinal microflora ferments prebiotics. Prebiotics saccharide belonging to the inulin group, GlcpFruf [α-D-fructopyranosyl-(β-D-fructofuranosyl) n-1-D-fructofuranose], Frup-Fruf [β-D-fructopyranosyl-(β-D-fructofuranosyl) n-1-D-fructofuranose] where n = 10–60; fructooligosaccharides, FOS (apart from oligofructose with n = 2–9 and, eventually, D-glucose, D-xylose, D-galactose and mannose residues) and non-digestible sugars, NDO. The oligofructose is present in wheat, onion, garlic, endive, leeks, asparagus, and others. In breast milk, the levels of oligosaccharides are relatively high, from 3 to 6 g/L; bifidobacterium selectively digests FOS and NDO and stimulates the development of colonies. Babies fed by the formula, that is, that refers to prebiotics, promote the development of probiotic bacteria, which inhibit the growth of pathogenic bacteria, the bacterial digestion of nutrients, especially proteins, and the decrease of their allergenicity. Probiotic bacteria synthesize cytoprotective short-chain acids, B and K vitamins, and polyamines and degrade the fiber by improving the action of the colon and by increasing the volume of feces [2].

Arabinoxylans (AX) (sometimes called pentosans) are the main non-cellulosic polysaccharides that are abundant in cereals. In the wheat and rye, they differ in their solubility in water, being some soluble and some others insoluble. Water-soluble and water-insoluble fractions are separated through stirring and from a centrifugation. These compounds have healthy benefits. The health benefits are prebiotic effect together with their ability antioxidant. This properties makes these are prevention agents of the colon cancer disease. At the metabolic level, the AX control glycemic and cholesterol levels are also immune regulator agents. During the later years, the research and interest of the AX have gathered considerable attention.

Arabinoxylans are composed of a chain, β(1–4) of xylose units, with variable branches of arabinose in (1–2) or (1–3) [3]. The solubility is modified according to the degree of substitution. The smaller the number of lateral branches, the greater the insolubility of the AX and the possibility of generating insoluble complexes [4]. The AX form viscous solutions with pseudoplastic behavior, and in the presence of certain oxidizing agents, the gels are much stronger and stable. This involves the action of ferulic acid in the formation of cross-links between the AX chains [5]. The texture of the bread increases during the cooking due to the cross-linking of the AX, whose natural levels in the wheat flour determine the quality of the bread, in the functional and nutritious properties [6].

2. Mechanism in the germination of cereals

Cereal beta-amylase is best known for the vital role it plays in the release of easily fermentable sugars from corn starch cereals to feed the production of alcohol by yeasts in beer brewing investigated because of its importance in this economically important brewing industry. Beta-amylases cereals are also used in other products of the food industry and in the analysis of starch. They are markers in the evaluation of cereals and in grain development studies. The enzymes β-amylases of cereals have been little studied at the physiological level.

In recent times, it was discovered that there are two categories of β-amylases, according to their pattern of tissue and taxonomic and physiological development. The classical β-amylases are present in the endosperm of the cereals of the family Triticeae of the subfamily Festucoideae of the grasses and the high levels of enzymatic activity, while the others are distributed in all tissues of the cereal but show less activity. The physiological phenomenology and the use of grain beta-amylases are discussed in relation to these two categories of enzymes [7]. The life cycle of a grain of cereal is divided into stages of development and germination, divided by a latency period [8]. During germination, much of the amino acid supply necessary for the growth of emerging seedlings comes from the degradation of the storage proteins of the seeds. The latter are synthesized during the maturation of the seeds and are deposited in specialized vacuoles. Little is known about the proteolytic enzymes in germinated rye. In contrast, the proteases that appear during the germination of barley [9–11] and wheat [12] are well studied. Using a non-denaturing electrophoretic system with protein substrate incorporated in the gel, [13] detected seven proteolytic bands in green barley malt. Five of these bands were maximally active at pH 3.8. Up to 20 proteolytic bands could be detected in extracts of germinated wheat grains [14]. The information shows that the total proteolytic activity increases during germination [15] has been investigated the temporal pattern of appearance of proteinases during germination, the sensitivities to specific inhibitors, and the location of proteinases and separated 42 activities into the germinating grain of barley using two-dimensional (2-D) gel electrophoresis (IEF × PAGE). These substances were divided into five groups based on isoelectric point (pI) values, PAGE mobilities, and biochemical characteristic [16]. About two-thirds of these enzymes were cysteine proteinases. The barley cysteine proteinases apparently hydrolyze most of the hordes, which are the main storage proteins of the barley endosperm. The purification and characterization of some of these cysteine proteinases have been described in several reports [17–20]. In [21] initiated the characterization of proteases in malted rye using two different methods to analyze their activities. A qualitative 2D method was used to measure the heterogeneity of the proteases and a method with solutions of different substrates to measure the activities of the different proteinase groups. It was concluded that the selected Humbolt rye cultivar produces high levels of proteolytic activities [22]. It was germinated under optimal conditions of temperature, aeration, and humidity, and the proteolytic activities were monitored using capillary electrophoresis methods. The total proteolytic activity was significantly higher after soaking and increased during the first 3 days of germination, but not after that time. The hydrolyzing activity was maximal at pH 53.5 and 45–50°C. After grinding, there were marked increases in the levels of proteolytic activity. The use of specific inhibitors showed the presence of four kinds of proteinases. According to the precedents, it is evident that hydrolysis of storage proteins in rye during germination is probably due to cysteines and aspartic proteinases [23].

3. Arabinoxylans and viscosity in the process of malting

Rye contains considerably higher amounts of arabinoxylans (AX), especially water-extractable arabinoxylans (WEAX), than barley. The rye AX structure and its properties were described [24]. AX are cell wall non-starch polysaccharides and are constituted by a chain of β-1,4 units of xylopyranose. The α-arabinofuranose chain can bind to the C(O)-2 and/or C(O)-3 of the xylose residues. It contains some arabinose substituents that are esterified with ferulic acid in C(O)-5 [25]. The molecular masses, the degree and the configuration of the arabinose branching, and the degree of esterification with ferulic acid exert an important influence on the properties of the AX. Since WEAX can bind to many water molecules, it can change the viscosity of the solutions making them more consistent. Gels can be formed by oxidative cross-linking of AX macromolecules through ferulic acid residues. The AX of the rye grains, during the process of the germination, suffer an enzymatic degradation. The inner bonds of the main chain of the xylan are divided forming a greater quantity of shorter-chain AX molecules by the action of the endoxylanase enzymes. Xylose residues are released from the nonreducing end since the β-xylosidases, while keeping the number of macromolecules constant, slightly decrease the length of the chain. The enzyme α-L-arabinofuranosidase releases arabinose from macromolecules, whereas feruloyl esterase hydrolyzes the bond between ferulic acid and arabinose residues [26]. AX represent a large part of the rye's dietary fiber, a soluble dietary fiber with their known health benefits for the consumer [27, 28] described in two studies, the positive effects of AX on postprandial insulin response in healthy individuals and improved metabolic control in diabetes patients. AX breakdown products have also been shown to display prebiotic properties [29–32, 40]. Certain proteins restrict access to water and enzymes to the endosperm starch. This situation is reversed during germination, when proteins and other components of the cellular apparatus are degraded and allow the passage of enzymes to the cellular interior. In this way, the extraction of starch and other compounds is increased. In the brewing industry, it is associated with a high nitrogen level at low starch content. The total nitrogen content of the malts obtained varied when the germination conditions changed. The difference between the highest and lowest values measured was 0.2%. But changes in germination conditions do not justify changes in the measurements obtained. The protein content in barley and rye from the beginning of germination did not vary significantly [33]. Nitrogen-containing compounds, being insoluble, will not form part of the mass during processing. In addition, some fractions of the soluble nitrogen (SN), like enzymes and free amino nitrogen (FAN), play important roles in the mashing process and during the fermentation, respectively. It was studied that the variations in SN measured produced from rye malts were large with values ranging between 694 and 862 mg/100 g. The highest SN contents were found in samples germinated for 96 h at 18°C, 130 h at 15°C, and 144 h at 10°C. In the germination of barley, storage proteins are mobilized, and some structural proteins are degraded. An increase in soluble nitrogen is generated in the malt, due to the large amount of peptides and extractable proteins in water. Protease enzymes influence the content of nitrogen compounds during maceration, although the greater amount of SN is formed in the malting. In rye, the content of SN does not appear to be influenced by the action of the proteolytic enzymes of the germinated malt. Previous studies show that free amino nitrogen (FAN), as an essential nutrient, is relevant in the growth of yeasts in the initial stage of fermentation. It was observed that at longer germination times, the FAN contents increase [34, 35]. In the maceration process, the starch is degraded by two enzymes, α-amylase and β-amylase, which act mainly generating maltose. Then, the maltose is fermented by yeasts that generate ethanol as a primary product. An increase in enzymatic activity has been observed, proportional to the time of germination. An increase in the total content of AX in the final malt has been observed, probably due to the loss of other compounds in respiration and formation of rootlets. The results of this study show which quality parameter of rye malt can be optimized by varying germination conditions. One of the most important factors, in the use of malt in the brewing industry, is the viscosity due to high amounts of WEAX. When the malt is used as a raw material for functional foods, its nutritional properties are used as a function of the AX and its decomposition products. According to the selected germination parameters, it is possible to direct the breakdown of the AX in the malt.

4. Prebiotic effects of cereal arabinoxylans

As reported before, in grains of wheat and related cereals, dietary fiber is predominantly composed of arabinoxylans (AXs). These cell wall components typically consist of a linear backbone of β-(1–4)-D-xylopyranosyl units, which may be substituted by α-L-arabinofuranosyl units [36]. The structure and the physicochemical properties of AX from different layers in cereal kernels are very diverse. The cereal's arabinoxylans (AX) are the main dietary fibers in a balanced human diet. The physiological impact of AX consumption strongly depends on their structures and properties as different impacts on the microbial population and fermentation products in the intestinal tract of rats. The consumption of extractable arabinoxylan of wheat bran increases the mass of the cecal contents. It is known that the consumption of soluble fermentable fiber by rodents causes increase in the mass of the cecum content. Extensive fermentation of water-extractable arabinoxylan or arabinoxylan oligosaccharides reduce the pH, suppress relevant markers of the proteolytic breakdown, and induce a selective bifidogenic response [37].

5. Application of rye malt in the brewing industries

The malting technologies were optimized for barley. The new materials must then be compared with the malt to establish identity in the processes. Wheat malt is also very important in large quantities in the western world. The contents of extract in rye malts were also higher than in barley malts (usually >81% [38]. As rye does not have a shell, which represents approximately 10% of the barley's dry weight [32], the fermentation capacities in the rye parts (73 and 77%) are lower than in 80% barley [38] or wheat parts (>78%, [39]), because they contain lower soluble nitrogen compounds and WEAX. However, the content of SN in rye malt turns out to be higher than in barley malt [38]. A higher viscosity of rye malt is the main drawback when using the brewing industry. This impediment can be reversed by modifying the temperature and the germination period (10°C and h) [40].

6. Folate in germinated cereals

The effects of germination and subsequent oven-drying or freeze-drying on folate content in a number of wheat and rye cultivars were studied for producing folate-rich flour ingredients. As reported in previous studies [41–43], germination of wheat and rye resulted in a four- to sixfold higher folate content than untreated cereals, mainly due to an increase in 5-CH3-H4 folate (w4-fold). The increased folate content has been attributed to de novo synthesis of folate being accelerated because of increased demand for methyl groups (one carbon unit) during germination [43]. Oven-drying (50°C) did not significantly affect the folate content, which is in agreement with our previous findings that oven-drying of germinated grains is suitable for the production of folate-rich candidate ingredients [44]. It has recently reported that by addition of germinated wheat flour to native wheat flour, bread with a 65% higher folate content compared with conventional Egyptian baladi bread could be prepared [45]. Germinated cereal grains and flour could also serve as functional ingredients for the European bread-baking industry. Here, the folate content in the rye cultivars studied was approximately 25% higher than that in the wheat cultivars. This confirms findings by [46] that rye flour contains more folate than flour from other cereals such as wheat, triticale, barley, and oats. The folate content quantified in the four Swedish wheat cultivars tested (23–33 mg/100 g dry matter DM) is also similar to HPLC data (34–40 mg/100 g DM) reported for four Polish wheat cultivars [48]. In comparison to data from microbiological assays [47–49], the mean folate content in untreated wheat and rye cultivars was 30–40% lower in this study. HPLC methods generally tend to provide folate values which are around

30% lower than data from microbiological assays [46–48]. Furthermore, the sum of folate content was underestimated by lack of quantification of 5-HCO-H4 folate and 10-formyl-dihydrofolate. Other groups also reported difficulties in quantification of 5-HCO-H4 folate in cereal foods using HPLC [45–47]. Up to 40% 5-HCO-H4 folate was reported in untreated wheat and rye [45–48], and the content was not significantly affected by germination of rye [49]. Also 7–13% of 10-formyl-dihydrofolate was found in untreated and germinated rye cultivars by [49, 50], while [51] did not detect this form. This can partly explain the discrepancy between our results and data reported in the literature. The mean folate content in six rye and four wheat cultivars ranges from 23 to 39 mg/100 g DM, being approximately 25% higher in rye than in wheat. The folate content in both cereals by four- to sixfold increases in germination subsequent oven-drying, which is required for milling of germinated grains, does not affect the folate content. Germinated kernels and their flours are ingredients with increased folate content for use in bakery products [52].

7. Prevent the bitterness of whole grain flour

The whole grain rye is beneficial for health. However it may be bitter. The impact of nonvolatile chemical compounds on the bitter taste of rye was analyzed by the aid of enzymatic hydrolysis, releasing potentially flavor-active compounds from the rye matrix. Water suspension of whole grain rye flour was treated with hydrolytic enzymes, after portions of the rye suspensions were baked into crackers and assessed for their sensory profile as well as solubilized hydrolysis products. Heat treatment reduced the perceived bitterness. The treatment with enzyme preparation with high protease activity increased the bitterness of rye and also wheat flour both as suspension and as crackers. Other enzymes tested (with high polygalacturonase, endo-glucanase, xylanase, or amyloglucosidase activity) had no significant impact on the perceived bitterness. Thus, small molecular weight peptides were considered to be a significant contributor to the bitter note of rye [53].

8. Relationship between the consumption of wholegrain cereals and incidence of lifestyle-related diseases

The incidence of diseases originating from the current lifestyle, such as cardiovascular diseases (CVD), certain types of cancer, and type II diabetes [54–62], is diminished by the consumption of whole grains [63, 65, 66]. Although importance is given to fibers in food, all the mechanisms responsible for this behavior are still not fully understood. Metabolomics was used for this study, which is a research that determines, in this case in plasma, the set of metabolites that is produced in the ingestion of a certain substance by means of instrumental analysis, for example, proton nuclear magnetic resonance spectroscopy ^{1}H NMR, chromatography, and mass spectrometry; these results are processed by means of a statistical analysis through software with the aim of understanding the endogenous biochemistry that is produced. When high-fat diets are consumed, apolipoprotein profiles of pigs are similar to humans, and these animals suffer atherosclerotic lesions similar to those that arise in humans [64]. Therefore, the pig is a suitable model to study the hypocholesterolemic effects of cereals. Barley and oats have been studied because of their high β-glucan content [60, 61, 67, 68]. However, there were no studies on whole meal rye breads; for this reason, using metabolomics as a tool, the positive effects of these breads were studied and compared with wheat breads in hypercholesterolemic pigs. The pigs were fed with rye-based ($n = 9$) or wheat ($n = 8$) bread with high-fat content and similar levels of dietary fiber for 9–10 weeks. Fasting plasma samples were collected 2 days before and after 8 and 12 days of consuming the experimental diets, while the postprandial samples were taken after 58–67 days, and the spectra of the ^{1}H NMR samples were made. The main component of the analysis (PCA) in the ^{1}H-NMR spectra of the plasma samples revealed a clear separation in the metabolite profiles of the plasma samples of the integral rye diet with respect to the samples of the diets of non-integral wheat both on day 8 and day 12 and at slaughter. In order to determine the differences in the metabolites of the two diets, a discriminant regression analysis by partial least squares (PLS-DA) was performed. On both day 8 and slaughter, an increase in the spectral intensities of the signals was observed at 3.29 ppm of the plasma samples of the animals that consumed rye compared with the samples of the animals that consumed the non-integral wheat, which is consistent with a chemical change for the different N(CH3)3 groups [64]. It has been demonstrated using liquid chromatography with LC-MS mass spectrometry detector that this response should be attributed to betaine [65]. We can affirm that the intake of a diet with high-fiber rye breads in hypercholesterolemic pigs increases betaine, which can be considered a biomarker because it is present in all the samples of the animals that consumed this diet with rye, which is not verified whether this biomarker can be used in a mixed diet [69, 70] since it has been shown that betaine is absorbed and increases in serum concentrations [72, 74]. Betaine acts as a methyl donor in the reaction of betaine-homocysteine methyltransferase that converts homocysteine into methionine [71, 73–76]. This is beneficial since plasma homocysteine is a risk factor for CVD [77, 78]; this property can be expected that betaine is involved in the prevention of CVD through this route. In addition, the concentration of plasma betaine is inverse to the amount of the following parameters, non-HDL serum cholesterol, triglycerides, and percentage of body fat, waist circumference, and systolic and diastolic blood pressure [79]. This study demonstrates a relationship between the intake of rye bread and the concentration of plasma betaine. Furthermore, in the analysis of the regressions, contributions of this diet to the chain of fatty acids linked to lipoproteins were observed; this behavior is not clear; however it could lead to an alteration in the composition [80].

Another study compared whether the postprandial glucose and insulin responses to whole-grain rye bread are less than wheat bread, and these responses were observed in two different types of rye bread. Rye breads are based on wholemeal flour and are thus rich in dietary fiber (DF). The dietary fiber content of our rye is 15 ± 17%, arabinoxylans (8 ± 10%), beta-glucan (2 ± 3%), and cellulose (1 ± 3%) being the main chemical constituents [81, 82]. Mainly due to its high DF content, wholemeal rye bread may reduce the health risks associated with coronary heart disease [79] and colon, breast, and prostate cancer [83, 84]. Slowly digestible carbohydrates have been suggested to be nutritionally most desirable, improving metabolic variables not only in diabetes and hyperlipidemia but also in healthy subjects [82, 83]. Although the glycemic index (GI) is a criticized concept [84], it is a widely used method for classification of different foods according to their effect on postprandial glucose levels. It has been analyzed that the glucose and insulin responses of different rye breads and other rye products have been reported to be variably lower than those of wheat bread [85]. The lowest GI values (66–80) have been reported for pumpernickel-type breads containing intact kernels [86–89]. There is a consensus that intact botanical structure protects the encapsulated starch of the kernel against the hydrolysis [90, 91]. The amount of whole kernels in the bread has been concluded to be more effective in reducing the glucose and insulin responses than the high-fiber content as such [86–88, 92]. The GI of food is generally to increase by the heat processing. However, there is exception to this rule, that is, pasta manufacture and most pasta products having a GI value of 50 ± 70 [93]. Low temperature and long-time baking may slow the digestion of bread by increasing the retrogradation of amylose and hence the amount of resistant starch (RS) in the product [94]. RS passes the small intestine without digestion and is available as energy only after colon fermentation. Rye bread contains organic acids and their salts; the latter are supposed to lower postprandial glucose and insulin responses [95–98] either by interfering the action of hydrolytic enzymes in the small intestine or by delaying gastric emptying [94]. The majority of the studies concerning glycemic responses of rye bread has been conducted in diabetic patients. In this study it was determined in healthy subjects whether the postprandial glucose and insulin responses to rye bread (whole kernel bread) are lower than those to wheat bread. Furthermore it was evaluated out if various types of rye breads give different glucose and insulin responses (wholemeal crispbread vs. wholemeal bread).

Standardized breads through an in vitro analysis of the hydrolysis rate of starch with a content of 43 ± 61 g of available carbohydrates, were consumed at a breakfast by 20 subjects (10 women and 10 men) with normal glucose tolerance. Eight samples of blood were taken from the subjects, postprandial for a period of 3 h. Eight samples of blood were taken from the subjects, postprandial for a period of 3 h. The results of the plasma insulin of the samples of the subjects who consumed whole wheat rye bread were lower than the samples of the subjects who consumed the wheat bread (45 min $P = 0.025$, 60 min $P = 0.002$, 90 min $P = 0.0004$, 120 min $P = 0.050$, 150 min $P = 0.033$); however there was no difference in glucose responses. We can conclude that wheat bread produces a greater postprandial insulin response than whole grain rye bread, but there is no difference in glucose response [99].

9. Conclusions

It is necessary to understand the impact of enzymes in AX and the behavior of rye proteins during the malting process. Although in the case of wheat this is studied with more depth, there is still no mass commercialization of wheat bread products. The malted and unmalted whole grain of rye and wheat contains fibers that are beneficial to prevent noncommunicable diseases. In malted flour compared to unmalted flour, the amount of soluble fibers increases. However, it is necessary to conduct research with cereals from different countries and compare the composition of these functional flours to apply them to different food products.

Acknowledgements

The authors gratefully acknowledge the Morón University for providing financial support.

Author details

María Laura Gómez Castro[1], Claudia Cecilia Larregain[1], Ethel Noemi Coscarello[1*] and Roberto Jorge Aguerre[1,2]

1 Laboratorio de Agroalimentos, Universidad de Morón, Buenos Aires, Argentina

2 Consejo Nacional de Investigaciones Científicas y Técnicas (Conicet), Buenos Aires, Argentina

*Address all correspondence to: ecoscarello@hotmail.com

References

[1] Diet Nutrition and the prevention of chronic diseases. Report of a Joint Food and Agriculture Organization of the United Nations (FAO) and World Health Organization WHO/FAO Expert Consultation GENOVE; 2004

[2] Morrison WR. Plant lipids in Research in Food Science and Nutrition. Dublin: Boole Press; 1984;**5**:247-260

[3] Lim S, Seib PA. Location of phosphate esters in a wheat starch phosphate by 31-P MNR spectroscopy. Cereal Chemistry. 1993;**70**(2):145-172

[4] Nakamura Y, Kubo A, Shimamune T, Matsuda T, Harada K, Satoh H. Correlation between activities of starch debranching enzyme and α-polyglucan structure in endosperms of sugary-1 mutants of rice. The Plant Journal. 1997;**12**(1):143-153. DOI: 10.1046/j.1365-313x.1997.12010143.x

[5] Colleoni C, Dauvillée D, Mouille G, Buléon A, Gallant D, Bouchet B, et al. Genetic and biochemical evidence for the involvement of α-1,4 glucanotransferases in amylopectin synthesis. Plant Physiology. 1999;**120**:993-1003

[6] Tomasik P. Chemical and Functional Properties of Food Saccharides. Boca Raton: CRC Press; 2004. ISBN 0-8493-1486-0

[7] Ziegler P. Cereal beta-amylases. Journal of Cereal Science. 1999;**29**: 195-204. DOI: 10.1006/jcrs.1998.0238

[8] Rock CD, Quatrano RS. The role of hormones during seed development. In: Plant Hormones. USA: Springer; 1995

[9] Wrobel R, Jones BL. Appearance of endoproteolytic enzymes during the germination of barley. Plant Physiology, American Society of Plant Biologists. 1992;**100**:1508-1516

[10] Zhang N, Jones BL. Development of proteolytic activities during barley malting and their localization in the green malt kernel. Journal of Cereal Science. 1995b;**22**:147-155

[11] Dominguez F, Cejudo FJ. Pattern of endoproteolysis following wheat grain germination. Physiologia Plantarum. Wiley; 1995;**95**(2):253-259

[12] Wrobel R, Jones BL. Electrophoretic study of substrate and pH dependence of endoproteolytic enzymes in green malt. Journal of the Institute of Brewing. 1992;**98**:471-178

[13] Koehler S, Ho T-H. Purification and characterization of gibberellic acid-induced cysteine endoproteases in barley aleurone layers. Plant Physiology. 1988;**87**:95-103

[14] Poulle M, Jones BL. A proteinase from germinating barley. I. Purification and some physical properties of a 30 kDa cysteine endoproteinase from green malt. Plant Physiology. 1988;**88**:1454-1460

[15] Phillips HA, Wallace W. Cysteine endopeptidase from barley malt which degrades hordein. Phytochemistry. 1989;**28**:3285-3290

[16] Zhang N, Jones BL. Purification and partial characterization of a 31-kDa cysteine endopeptidase from germinated barley. Planta. 1996;**199**:565-572

[17] Brijs K, Trogh I, Jones BL, Delcour JA. Proteolytic enzymes in germinating rye grains. Cereal Chemistry. 2002;**79**(3):423-428

[18] Brijs K, Bleukx W, Delcour JA. Proteolytic activities in dormant rye (*Secale cereale* L.) grain. Journal of Agricultural and Food Chemistry. 1999;**47**:3572-3578

[19] Brijs K, Trogh I, Jones BL, Delcour JA. Proteolytic enzymes in germinating rye grains. Journal of Cereal Science. 1999;**29**:195-204. Available online at http://www.idealibrary.com. DOI: 10.1094/CCHEM.2002.79.3.423

[20] Dunaevsky YE, Sarbakanova ST, Belozersky MA. Wheat carboxypeptidase and joint action on gliadin of proteases from dry and germinating seeds. Journal of Experimental Botany. 1989;**40**:1323-1329

[21] Enari T-M, Sopanen T. Mobilisation of endospermal reserves during the germination of barley. Journal of the Institute of Brewing. 1986;**92**:25-31

[22] Mikola M, Jones BL. Electrophoretic and "in solution" analysis of endoproteinases extracted from germinated oats. Journal of Cereal Science. 2000;**31**:15-23

[23] Rastogi V, Oaks A. Hydrolysis of storage proteins in barley endosperms: Analysis of soluble products. Plant Physiology. 1986;**81**:901-906

[24] Shutov AD, Vaintraub IA. Degradation of storage proteins in germinating seeds. Phytochemistry. 1987;**26**:1557-1566

[25] Vinkx CJA, Delcour JA. Rye (*Secale cereale* L.) arabinoxylans a critical review. Journal of Cereal Science. 1996;**24**:1-14

[26] Bengtsson S, Andersson R, Westerlund E, Åman P. Content, structure and viscosity of soluble arabinoxylans in rye grain from several countries. Journal of the Science of Food and Agriculture. 1992;**58**:331-337

[27] Lu ZX, Walker KZ, Muir JG, Mascara T, O'Dea K. Arabinoxylan fibre, a byproduct of wheat flour processing, reduces the postprandial glucose response in normoglycemic subjects. American Journal of Clinical Nutrition. 2000;**71**:1123-1228

[28] Lu ZX, Walker KZ, Muir JG, O'Dea K. Arabinoxylan fibre improves metabolic control in people with type II diabetes. European Journal of Clinical Nutrition. 2004;**58**:621-628

[29] Grootaert C, Delcour JA, Courtin CM, Broekaert WF, Verstraete W, Van de Wiele T. Microbial metabolism and prebiotic potency of arabinoxylan oligosaccharides in the human intestine. Trends in Food Science and Technology. 2007;**18**:64-71

[30] Cloetens L, DePreter V, Swennen K, Broekaert WF, Courtin CM, Delcour JA, et al. Dose response effect of arabinoxylo oligosaccharides on gastrointestinal motility and oncolonic bacterial metabolism in healthy volunteers. Journal of the American College of Nutrition. 2008;**27**(4):512-518

[31] Courtin CM, Swennen K, Broekaert WF, Swennen Q, Buyse J, Decuypere E, et al. Effects of dietary inclusion of xylo oligosaccharides, arabinoxylo oligosaccharides and soluble arabinoxylan on the microbial composition of cecal contents of chickens. Journal of the Science of Food and Agriculture. 2008;**88**:2517-2522

[32] VanCraeyveld V, Swennen K, Dornez E, VandeWiele T, Marzorati M, Verstraete W, et al. Structurally different wheat-derived arabinoxylooligosaccharides have different prebiotic and fermentation properties in rats. Journal of Nutrition. 2008;**138**:2348-2355

[33] Briggs DE. Malt and Malting. London: Blackie Academic and Professional; 1998

[34] Jones BL. Endoproteases of barley and malt. Journal of Cereal Science. 2005;**42**:139-156

[35] Jones BL, Budde AD. How various malt endoproteinase classes affec twort

soluble protein levels. Journal of Cereal Science. 2005;**41**:95-106

[36] Izydorczyk MS, Biliaderi GC. Cereal arabinoxylans: Advances in estructure and physicochemical properties. Carbohydrate Polymers. 1995;**28**:33-48

[37] Damen B, Verspreet J, Pollet A, Broekaert WF, Delcour JA, Courtin CM. Prebiotic effects and intestinal fermentation of cereal arabinoxylans and arabinoxylan oligosaccharides in rats depend strongly on their structural properties and joint presence. Molecular Nutrition & Food Research. 2011;**55**:1862-1874

[38] Back W, et al. Ausgewählte Kapitelder Brauerei Technologie. (Selected chapters of brewingtechnology). Nürnberg: Fachverlag Hans Carl GmbH; 2005

[39] MEBAK. Brautechnische Analysemethoden (in German). 3rd ed. Freising-Weihenstepha: MEB AK; 1997

[40] Hübner F, Schehl BD, Gebruersc K, Courtin CM, Delcour JA, Arendt EK. Influence of germination time and temperature on the properties of rye malt and rye malt based worts. Journal of Cereal Science. 2010;**52**:72-79

[41] Kariluoto S, Liukkonen K, Myllymäki O, Vahteristo L, Kaukovirta-Norja A, Piironen V. Effect of germination and thermal treatments on folates in rye. Journal of Agricultural and Food Chemistry. 2006;**54**:9522-9528

[42] Koehler P, Hartmann G, Wieser H, Rychlik M. Changes of folates, dietary fiber, and proteins in wheat as affected by germination. Journal of Agricultural and Food Chemistry. 2007;**55**:4678-4683

[43] Hefni M, Witthöft C. Increasing the folate content in Egyptian baladi bread using germinated wheat flour. LWT—Food Science and Technology. 2011;**44**:706-712

[44] Jabrin S, Ravanel S, Gambonnet B, Douce R, Rebeille F. One-carbon metabolism in plants. Regulation of tetrahydrofolate synthesis during germination and seedling development. Journal of Plant Physiology. 2003;**131**:1431-1439

[45] Gujska E, Kuncewicz A. Determination of folate in some cereals and commercial cereal-grain products consumed in Poland using trienzyme extraction and high-performance liquid chromatography methods. European Food Research and Technology. 2005;**221**:208-213

[46] Konings EJ. A validated liquid chromatographic method for determining folates in vegetables, milk powder, liver, and flour. Journal of AOAC International. 1999;**82**:119-127

[47] Kariluoto S, Vahteristo L, Piironen V. Applicability of microbiological assay and affinity chromatography purification followed by high-performance liquid chromatography (HPLC) in studying folate contents in rye. Journal of the Science of Food and Agriculture. 2001;**81**:938-942

[48] Konings EJ, Roomans HH, Dorant E, Goldbohm RA, Saris WH, van den Brandt PA. Folate intake of the Dutch population according to newly established liquid chromatography data for foods. American Journal of Clinical Nutrition. 2001;**73**:765-776

[49] Nyström L, Lampi A-M, Andersson AAM, Kamal-Eldin A, Gebruers K, Courtin CM, et al. Phytochemicals and dietary fibre components in rye varieties in the HEALTHGRAIN diversity screen. Journal of Agriculture and Food Chemistry. 2008;**56**:9758-9766

[50] Piironen V, Edelmann M, Kariluoto S, Bedo Z. Folate in wheat genotypes in the HEALTHGRAIN diversity screen. Journal of Agriculture and Food Chemistry. 2008;**56**:9726-9731

[51] Pfeiffer CM, Rogers LM, Gregory JF. Determination of folate in cereal-grain food products using trienzyme extraction and combined affinity and reversed-phase liquid chromatography. Journal of Agricultural and Food Chemistry. 1997;**45**:407-413

[52] Hefni M, Witthöft CM. Effect of germination and subsequent oven-drying on folate content in different wheat and rye cultivars. Journal of Cereal Science. 2012;**56**:374-378. DOI: 10.1016/j.jcs.2012.03.009

[53] Heiniö R-L, Nordlund E, Poutanen K, Buchert J. Use of enzymes to elucidate the factors contributing to bitterness in rye flavour. Food research international. 2012;**45**(1):31-38

[54] Dixon JL, Stoops JD, Parker JL, Laughlin MH, et al. Dyslipidemia and vascular dysfunction in diabetic pigs fed an atherogenic diet. Arteriosclerosis, Thrombosis, and Vascular Biology. 1999;**19**:2981-2992

[55] Mellen PB, Walsh TF, Herrington DM. Whole grain intake and cardiovascular disease: A meta-analysis. Nutrition, Metabolism, and Cardiovascular Diseases. 2008;**18**:283-290

[56] Levi F, Pasche C, Lucchini F, Chatenoud L, et al. Refined and whole grain cereals and the risk of oral, oesophageal and laryngeal cancer. European Journal of Clinical Nutrition. 2000;**54**:487-489

[57] Slavin JL, Jacobs D, Marquart L, Wiemer K. The role of whole grains in disease prevention. Journal of the American Dietetic Association. 2001;**101**:780-785

[58] Truswell AS. Cereal grains and coronary heart disease. European Journal of Clinical Nutrition. 2002;**56**:1-14

[59] McKeown NM, Meigs JB, Liu S, Wilson PW, Jacques PF. Whole-grain intake is favorably associated with metabolic risk factors for type 2 diabetes and cardiovascular disease in the Framingham Offspring Study. The American Journal of Clinical Nutrition. 2002;**76**:390-398

[60] Pereira MA, Jacobs DR, Pins JJ Jr, Raatz SK, et al. Effect of whole grains on insulin sensitivity in overweight hyperinsulinemic adults. The American Journal of Clinical Nutrition. 2002;**75**:848-855

[61] Fung TT, Hu FB, Pereira MA, Liu S, et al. Wholegrain intake and the risk of type 2 diabetes: A prospective study in men. The American Journal of Clinical Nutrition. 2002;**76**:535-540

[62] KEB K, Canibe N. In: Lairon D, editor. Proceedings of Cost 92 Workshop COST 92. Metabolic and Physiological Aspects of Dietary Fibre in Foods. Luxembourg: Commission of the European Communities; 1993. pp. 123-130

[63] Terpstra AHM, Lapre JA, de Vries HT, Beynen AC. Transiency of the different cholesterolaemic responses to dietary cellulose and psyllium in pigs and two strains of hamsters. Journal of Animal Physiology and Animal Nutrition. 2000;**84**:178-191

[64] Ribsin CM, Keenan JM, Jacobs DR, Elmer PJ, et al. Oat products and lipid lowering: A metaanalysis. JAMA. 1992;**267**:3317-3325

[65] Aman P. Cholesterol-lowering effects of barley dietary fibre in humans: Scientific support for a generic health claim. Scandinavian Journal of Food and Nutrition. 2006;**50**:173-176

[66] Lærke HN, Pedersen C, Mortensen MA, Theil PK, et al. Rye bread reduces plasma cholesterol levels in hypercholesterolaemic pigs when

compared to wheat at similar dietary fibre level. Journal of the Science of Food and Agriculture. 2008;**88**:1385-1393

[67] Lindon JC, Nicholson JK, Everett JR. NMR spectroscopy of biofluids. Annual Reports on NMR Spectroscopy. 1999;**38**:1-88

[68] Bertram HC, Bach Knudsen KE, Serena A, Malmendal A, et al. NMR-based metabonomic studies reveal changes in the biochemical profile of plasma and urine from pigs fed high-fibre rye bread. The British Journal of Nutrition. 2006;**95**:955-962

[69] Chen Y, Ross AB, Aman P, Kamal-Eldin A. Alkylresorcinols as markers of whole grain wheat and rice in cereal products. Journal of Agricultural and Food Chemistry. 2004;**52**:8242-8246

[70] Linko-Parvinen AM, Landberg R, Tikkarren MJ, Adlercreutz H, Penalvo JL. Alkylresorcinols from wholegrain wheat and rye are transported in human plasma lipoproteins. The Journal of Nutrition. 2007;**137**:1137-1142

[71] Frontiera MS, Stabler SP, Kolhouse JF, Allen RH. Regulation of methionine metabolism: Effects of nitrous oxide and excess dietary methionine. The Journal of Nutritional Biochemistry. 1994;**5**:28-38

[72] Schwab U, Törrönen A, Toppinen L, Alfthan G, et al. Betaine supplementation decreases plasma homocysteine concentrations but does not affect body weight, body composition, or resting energy expenditure in human subjects. The American Journal of Clinical Nutrition. 2002;**76**:961-967

[73] Schwahn BC, Hafner D, Hohlfeld T, Balkenhol N, et al. Pharmacokinetics of oral betaine in healthy subjects and patients with homocystinuria. British Journal of Clinical Pharmacology. 2003;**55**:6-13

[74] McGregor DO, Dellow WJ, Robson RA, Lever M, et al. Betaine supplementation decreases post-methionine hyperhomocysteinemia in chronic renal failure. Kidney International. 2002;**61**:1040-1046

[75] Delgado-Reyes CV, Garrow TA. High sodium chloride intake decreases betaine-homocysteine S-methyltransferase expression in guinea pig liver and kidney. American Journal of Physiology. Regulatory, Integrative and Comparative Physiology. 2005;**288**:182-187

[76] Holm PI, Bleie O, Ueland PM, Lien EA, et al. Betaine as determinant of postmethionine load total plasma homocysteine before and after B-vitamin supplementation. Arteriosclerosis, Thrombosis, and Vascular Biology. 2004;**24**:301-307

[77] Holm PI, Hustad S, Ueland PM, Vollset SE, et al. Modulation of the homocysteine-betaine relationship by methylenetetrahydrofolate reductase 677 C->T genotypes and B-vitamin status in a large scale epidemiological study. The Journal of Clinical Endocrinology and Metabolism. 2007;**92**:1535-1541

[78] Craig SAS. Betaine in human nutrition. The American Journal of Clinical Nutrition. 2004;**80**:539-549

[79] Konstantinova SV, Tell GS, Vollset SE, Nygård O, Bleie Ø. Divergent associations of plasma choline and betaine with components of metabolic syndrome in middle age and elderly men and women. The Journal of Nutrition. 2008;**138**:914-920

[80] Bertram HC, Duarte IF, Gil AM, Bach Knudsen KE, Lærke HN. Metabolic profiling of liver from hypercholesterolemic pigs fed rye or wheat fibres and of liver from normal pigs fed a standard diet—A high-resolution magic angle spinning 1H

NMR spectroscopic study. Analytical Chemistry. 2007;**79**:168-175. 1062 i 2009 WILEY-VCH Verlag GmbH & Co. KGaA, Weinheim www.mnf-journal

[81] HaÈrkoÈnen H, Pessa E, Suortti T, Poutanen K. Distribution and some properties of cell wall polysaccharides in rye milling fractions. Journal of Cereal Science. 1997;**26**:95-104

[82] Aman P, Nilsson M, Andersson R. Chemical composition of rye grain. In: Poutanen K, Autio K, editors. International Rye Symposium: Technology and Products. VTT Symposium 161. Espoo: VTT Offset Press; 1995. pp. 33-38

[83] Pietinen P, Rimm EB, Korhonen P, Hartman AM, Willett WC, Albanes D, et al. Intake of dietary fiber and risk of coronary heart disease in a cohort of Finnish men. The alpha-tocopherol, beta-carotene cancer prevention study. Circulation. 1996;**94**:2720-2727

[84] Consensus meeting, 1997. Consensus meeting on cereals, fibre and colorectal and breast cancer: ECP consensus panel on cereals and cancer. European Journal of Cancer Prevention. 1997;**6**:512-514

[85] Zhang J-X, Hallmans G, LandstroÈm M, Bergh A, Damber J-E, A Ê man P, et al. Soy and rye diets inhibit the development of dunning R3327 prostatic adenocarcinoma in rats. Cancer Letters. 1997;**114**:313-314

[86] BjoÈrck I, Granfeldt Y, Liljeberg H, Tovar J, Asp NG. Food properties affecting the digestion and absorption of carbohydrates. The American Journal of Clinical Nutrition. 1994;**59**:699-705

[87] Englyst HN, Hudson GJ. Starch and health. In: Frazier PJ, Donald AM, Richmon P, editors. Starch: Structure and Functionality. Cambridge: Royal Society of Chemistry; 1997. pp. 9-21

[88] TMS W. The glycemic index: Flogging a dead horse. Diabetes Care. 1997;**20**:452-456

[89] Foster-Powell K, Brand Miller J. International tables of glycemic index. The American Journal of Clinical Nutrition. 1995;**62**(Suppl):S871-S893

[90] Jenkins DJA, Wolever TMS, Kalmusky J, Giudici S, Giordano C, Wong GS, et al. Low glycemic index carbohydrate foods in the management of hyperlipidemia. The American Journal of Clinical Nutrition. 1985;**42**:604-617

[91] Jenkins DJA, Wolever TMS, Jenkins AL, Giordano C, Giudici S, Thompson LU, et al. Low glycemic response to traditionally processed wheat and rye products: Bulgur and pumpernickel bread. The American Journal of Clinical Nutrition. 1986;**43**:516-520

[92] Wolever TMS, Jenkins DJA, Josse RG, Wong G, Lee R. The glycemic index: Similarity of values derived in insulin-dependent and non-insulin-dependent diabetic patients. Journal of the American College of Nutrition. 1987;**6**:295-305

[93] Wolever TMS, Katzman-Relle L, Jenkins AL, Vuksan V, Josse RG, Jenkins DJA. Glycaemic index of 102 complex carbohydrate foods in patients with diabetes. Nutrition Research. 1994;**14**:651-669

[94] Liljeberg H, Granfeldt Y, BjoÈrck I. Metabolic responses to starch in bread containing intact kernels versus milled flour. European Journal of Clinical Nutrition. 1992;**46**:561-575

[95] BjoÈrck I, Liljeberg H. Dietary fiber, resistant starch and other food factors as moderators of acute glycaemia and second-meal tolerance: Review. In: Guillon F, Abraham G, Amado R, Andersson H, Asp NG, Bach Knudsen KE, Champ M, Robertson J, editors.

Plant Polysaccharides in Human Nutrition: Structure, Function, Digestive Fate & Metabolic Effects. Nantes: Imprimerie Parentheses; 1997. pp. 79-85

[96] Liljeberg H, BjoÈrck I. Bioavailability of starch in bread products. Postprandial glucose and insulin responses in healthy subjects and in vitro resistant starch content. European Journal of Clinical Nutrition. 1994;**48**:151-163

[97] Liljeberg HGM, LoÈnner CH, BjoÈrck IME. Sourdough fermentation or addition of organic acids or corresponding salts to bread improves nutritional properties of starch in healthy humans. The Journal of Nutrition. 1995;**125**:1503-1511

[98] Liljeberg HGM, BjoÈrck IME. Delayed gastric emptying rate as a potential mechanism for lowered glycemia after eating sourdough bread: Studies in humans and rats using test products with added organic acids or organic salt. The American Journal of Clinical Nutrition. 1996;**64**:886-893

[99] Bertram HC, Malmendal A, Nielsen NC, Straadt IK, Larsen T, Knudsen KEB, et al. NMR-based metabonomics reveals that plasma betaine increases upon intake of high-fiber rye buns in hypercholesterolemic pigs. Molecular Nutrition & Food Research. 2009;**53**:1055-1062. DOI: 10.1002/mnfr.200800344